A special message from Dr. John J. Woods:

We are subject to all sorts of threats. These threats are ever-growing and we must be ready to withstand them, prevail over them.

There is much to be learned in planning, stocking, training, and acquiring the skills of self-defense and protection. My advice: Begin now with this initiation into survival weaponry. It's a good place to start the process.

Gun up now. You can drink, eat, and sleep later.

– John J. Woods

BASIC PREPPING ESSENTIALS
—— WEAPONS ——

JOHN J. WOODS

A PERMUTED PRESS BOOK

ISBN: 978-1-68261-292-7

Basic Prepping Essentials: Weapons
© 2017 by John J. Woods
All Rights Reserved

Cover art by Christian Bentulan

No part of this book may be reproduced, stored in a retrieval system, or transmitted by any means without the written permission of the author and publisher.

Permuted Press, LLC
permutedpress.com

Published in the United States of America

CONTENTS

CHAPTER 1
GUN REVIEWS 1

CHAPTER 2
PREPPER AMMO AND CARTRIDGES 34

CHAPTER 3
AR RIFLE ACCESSORIES AND STRATEGIES 58

CHAPTER 4
SHOOTING GEAR AND TACTICS 76

CHAPTER 5
HANDGUN ACCESSORIES, GEAR AND TACTICS 103

CHAPTER 6
PREPPER FIREARMS AND SUPPORT GEAR 128

CHAPTER 7
PREPPER GEAR 162

CHAPTER 8
PREPPER STRATEGIES 179

CHAPTER 9
GUN AND GEAR MAINTENANCE 204

ABOUT THE AUTHOR 212

CHAPTER 1
GUN REVIEWS

300 AAC-BLACKOUT-WHISPER/HORNADY PROVES METTLE ON WHITETAILS

Honestly, you never know the true performance of a particular gun and ammo combination until it is tested in the field under real hunting conditions. So, in the back of my mind there was a bit of nervous anxious anticipation when I pulled the trigger on my Smith and Wesson MP15T with a 300 AAC/Blackout/Whisper upper unit on a white-tailed deer.

The Hornady ammo of course produced the harvest results with one clear, clean shot at 150 yards in dying light with a misting, foggy atmosphere covering the food plot. The Hornady Custom 300 Whisper (300 Blackout) load uses their 110-grain VMax polymer-tipped bullet. The ballistics are good for deer hunting, hogs, or varmints.

First let's clear up some technical stuff, aka marketing stuff. As per the title here you note the reference to the 300 AAC (Advanced Armament Corporation), the Blackout or BLK, and the 300 Whisper. For all intents and purposes they are the same.

The Smith and Wesson MP15T's upper unit's barrel is marked 300 Whisper dash 300 AAC Blackout. In the 2014 Hornady factory ammo catalog the load is listed as the 300 Whisper. I note in the 2015 catalog however, the "Whisper" namesake is gone in favor of just the 300 Blackout. Any user has to know they are the same in all practical aspects. Hopefully all the ammo makers marketing people will standardize these so people will not be further confused.

Years ago when I was hunting with Smith and Wesson along with several writers and editors, as sponsored by the great marketing group Blue Heron Communications out of Norman, Oklahoma, I was introduced for the first time to the 300 Whisper. We were hunting deer and hogs with .223s and 300 Whisper S&W AR rifles with EOTech electronic optical sights. That whole crew provided great instruction and range practice with the 300 before the hunts took place. Then we used Hornady and DRT ammo.

I came away from that hunt knowing I wanted to build an AR rifle for deer and hogs on the 300 Whisper concept. The more I researched the 300 Whisper the more I came to realize this was also the 300 AAC as well as the 300 Blackout, by name anyway.

As factory ammo came out, you could find it in all the menu varieties, but all performed (tasted) relatively the same. To date I have ammo from Hornady, PNW Arms, and Remington all marked somewhat differently. Hunting ammo for the 300 AAC-BLK-Whisper uses a variety of bullets ranging in weights from 110, 115, and 125 grains. The bullet choices I have in stock include the V-Max, the Nosler Ballistic Tip BT, and the CTFB (Closed Tip Flat Based). By now I am sure other loads are on the market.

This 300-series cartridge yields velocities in the range of 2,200 feet per second and 1,360 foot pounds of energy with a 125-grain bullet. There are also subsonic loads for use with a suppressor, which is becoming more popular all the time for varmint hunting applications. If you want a comprehensive review of this cartridge, check it out on *Wikipedia*.

The 300 AAC-Blackout has an interesting history of development too lengthy to relate here, so *Google* it to see the whole profile. In ammo terminology this 300 is the 7.62x35mm and was created to compete with Soviet AK cartridge in hopes the U.S. military might adopt it. So far that has not happened, and I seriously doubt it ever will.

If you are an AR man currently using one in .223/5.56 with multiple magazines, then you'll be happy to know the 300 AAC-Blackout-Whisper ammo will fit and function just fine in existing AR magazines for the .223. All you have to do is obtain a 300 AAC upper unit and install it on your standard AR lower. That's what I did.

In the hog rifle "build" I got the upper unit from Smith and Wesson to go on my SWMP-15T lower. It was an instantaneous and perfect match. This upper has a rail system for adding accessories with easy, secure attachments.

I added a Nite Hunter green light flashlight in their excellent mount for low light or night hog hunting. I locked down Midwest Industries push button sling attachment points front and rear to mate my Vero Vellini neoprene sling for comfortable carry. I installed a Hogue soft rubber AR grip. I also put on a CAA soft touch cheek rest cover over the top part of the adjustable AR tube buttstock.

Scopewise I obtained the excellent Leupold VX-R *HOG* model, 1.25-4x20 optic with the illuminated reticle. The reticle on this scope has a small circle in the center with hash range marks below. The 30mm tube adds extra light gathering capability to which I attested during my deer hunt. I did not even have to turn on the lighted reticle circle to paint the target. I also used the super one-piece AR Leupold mount for this scope. The scope came with a high-quality neoprene protective scope cover as well.

The only other thing I did to the rifle was add left, right, and lower AR rail covers to protect the hands from the abrasive edges of the rail slots. This keeps my woolen gloves from being worn out so fast, too. The Tapco covers are semi-soft and allow for a good non-slip grip for firm targeting.

Range sessions for my AR deer/hog rig proved promising as well. This rifle is a pleasure to shoot and the 300 Whisper-BLK cartridge generates little recoil and is mild in muzzle blast noise. I was easily producing 1-2 inch 100-yard groups with a selection of ammo. I had good confidence this package would deliver its load in the field right on target.

Finally getting back into a deer stand after Christmas, I was ready for action. I watched a nubbin buck feed in the oat-wheat food plot for nearly an hour, but no other action. Near 4:30 p.m. a small young forkhorn buck walked across the end of a shooting lane food plot.

All this time a fog was developing across the woods and a light misting sort-of rain was coming down with a slight breeze. At 5:10 p.m. a lone doe walked into the plot at the far end. I glassed and glassed to make sure this was not a button or spike buck. It was indeed a solo doe, which is unusual on our hunting property. Usually a doe will have at least one yearling or two in company, but not this time.

I waited for the doe to turn broadside, and in the process of moving my rifle up into position on the stand's shooting rail, the doe caught my movement. I froze and for several minutes she just watched me. Then she resumed feeding. A few moments later she offered the shot I was looking for, and the AR rifle-scope-ammo package delivered. One shot to the shoulder and the deer was instantly down. My anxiety dissipated like the fog.

The Hornady VMax bullet did its job, the Smith AR did its job, as did the Leupold scope under poor light conditions. Proof of its effectiveness is in my neighbor's freezer, a gift to help him feed his family of six.

The 300 Blackout is certainly worthy of consideration for a variety of hunting applications. The nice part is if you already own an AR platform, all you need to do is acquire another upper unit, install it and then customize it as you wish. There is no license, BATF 4473 registration or dealer transaction either as the serial number is on your lower already.

If you want or need another hunting option with more power for your AR or a new one, then give the 300 AAC-Blackout-Whisper a try. That is whatever you want to call it.

A PEEK AT THE PICO

The much-hyped Beretta Pico has finally arrived on dealers' shelves. Well, sort of. The local Academy store was to get some in according to a sales flyer that came in the local newspaper. However, the counter sales team at both area stores knew nothing about it, and the company's website did not list the Beretta Pico either.

My Beretta collector friend John called the store and was told they did not have any and did not know what the *Pico* even was. So, he went to the store the next day just to see and there it was in the glass display case. He was told they got in three and only had one left in stock.

He bought it immediately after a search for one of nearly two years. That was the timeframe announced long ago that the new pistol would be available. It could be that deliveries are finally being made in earnest. I've now seen them in shops.

If ever there was a gun hyped and promoted early on by the mainstream gun media, the Beretta Pico is it. I have witnessed this happen many times over my thirty-plus years as an outdoor writer. I remember reading one national review of a new Ruger handgun which implied it was already for sale at local dealers, but it was several years later before it became available. The backlash was severe.

Another time I read a story on a new Browning hunting rifle in a chambering that I had wanted for years. It was the .284 Winchester. It was even listed in their new catalog, but the rifle never appeared, ever.

So now, finally comes the Pico after much has been written about apparently writer-only samples. Those initial reports are several years old. Actually, most of the reviews have only been press-release-type material and not actual field tests by knowledgeable gun writers.

I never have known who was really at fault over these preemptive promotions of guns and hunting equipment, but I get a lot of inquiries about such when buyers cannot find the products actually for sale. It is frustrating, but just part of the process.

The good part is none of the delayed marketing phony-baloney has anything to do with the quality of the product. In this case, I can now attest that a Beretta is a Beretta no matter the final delivery date. This new Pico is no exception.

So, what exactly is a Pico anyway? Once you finally get to see and handle the new Pico you are probably going to think "this is a mini-Nano." Well, at first appearance that would not be a bad first impression. The Pico certainly has the look of the Nano, but in a downsized version.

The Beretta Pico is essentially a small, compact, semi-auto pistol designed and intended for concealed carry and personal protection. It is chambered for the .380 ACP and is thus primarily intended for private citizen use, not really meant for law enforcement except maybe as a backup gun, but certainly not for military use.

The Pico is double-action only, so there is no external hammer to manually cock. The pistol weighs in at 11.5 ounces with 5.1 inches in overall length. Its barrel is only 2.7 inches long. The sights are iron sights. The pistol's magazine holds six rounds plus one in the chamber.

The Pico's frame is made of a polymer to reduce weight. The frame grip is easy to secure and handle during firing sessions. The gun's slide is mostly made from stainless steel. The pistol's edges are smooth and rounded so as to reduce snagging on anything such as a holster or even pants pockets or coat jackets, etc.

Other features on the Pico include some ambidextrous features including the magazine release. As per usual Beretta pistol designs, this one is easy to take down as well. Just the turn of one screw allows the gun to be disassembled quickly. This makes internal cleaning very easy.

There are not a lot of aftermarket accessories for the Pico yet, but you can supposedly get alternately colored polymer frames from the Beretta factory website. I did this with my Nano, going to an olive-drab frame from the standard black. It's cool looking.

BASIC PREPPING ESSENTIALS: WEAPONS

Just like the Nano, the smaller Pico appears boxy in appearance and thus not exactly appealing at first glance. It grows on you, especially after you pick it up, handle it, break into a defensive draw stance, or try to stuff it in your pants pocket. You have to pick the gun up several times, get a good grip, wave it around some like you're playing a ghetto thug, and then align the sights. Very quickly the square design produces a smile if you spend any time with it at all.

The .380 ACP is a pleasant, low-recoil round to shoot. It is easy to control for repeated shots in the Beretta Pico. The .380 ammo choices are many, coming with a wide variety of bullet weights and types. The "standard" bullet weight in the .380 is the 95 grain. You can find factory ammo offerings with 50, 90, 91, 94, and 100-grain bullets. There are probably others as well. Several personal defense loads are available, too.

As with any semi-auto pistol, the owner-shooter has to find a load or two that fully functions well, cycles the gun's action, and puts bullets in the proper place. As I told a gent I was counseling recently at the shooting range, a self-defense pistol like the Pico is not intended to be shot at 50 yards, or even 20. These are 10-yard guns max, but more like 10 feet. They are designed for close quarters, up close and personal self-defense.

For many, the initial impression of the .380 ACP is that it is too light for practical self-defense, but it has long proven itself otherwise. Because of the concealment capability of the Pico it makes an excellent personal protection firearm when using a good load designed for self-defense.

As with all the Beretta guns I have ever used, including my very first semi-auto shotgun that my mom got me for college graduation, Berettas have just proven to be high-quality, well-made, highly functional, reliable guns. Otherwise why would the U.S. military pick the Model 92 as their primary sidearm?

If you are considering shopping for a small, compact semi-auto for personal protection, then put the Pico on your list to inspect.

Hopefully, you'll be able to find one finally at a dealer near you. If so, you better get it fast.

THE NEW GLOCK 9MM HANDGUN

It has been rumored for several years that Glock was working on a single stack, 9mm handgun for concealed carry. Working gun shows in our area for GlockPro of Mississippi, every event we fielded questions about when the new Glock was coming out. Despite the whirlwind of false announcements, mistaken sightings and other misinformation, the wait is over.

At the NRA Annual Meeting in Nashville, the Glock booth debuted the new Glock Model G43. This new Glock 43 is a slimline design for everyday concealed carry needs. The G43 is ultra-concealable, accurate, and comfortable for all shooters, regardless of hand size.

A true slimline pistol, the G43 frame is just over an inch wide and the slide width is only 0.87 inch. The overall length of the new pistol is 6.26 inches with a trigger distance of only 2.6 inches. This will make handling the new Glock ideal for shooters with smaller hands.

The single stack magazine will hold six rounds, again making the pistol's slim profile perfect for concealed carry. The pistol will be a good choice for both duty use and desirable for civilian shooters as well.

KEL-TEC'S KSG

There is nothing else like it on the market today. The look, design, feel, and features make the Kel-Tec KSG one unique firearm. One can maybe debate its principal utility, but not the functional alternatives it offers.

The KSG is a 12-gauge, pump-action shotgun of the classic bullpup design orientation. This means the pistol grip and the trigger are forward of the receiver action. Underneath the 18.5-inch barrel are twin parallel tubular magazines capable of holding fourteen 2¾-inch shotshells, or twelve 3-inch shells. The smoothbore is choked cylinder bore or open choke. It will handle slugs, buckshot, birdshot, or any combination.

The KSG's overall length is 26-inches with an empty weight of seven pounds. Loaded the weight climbs to roughly 8.5 pounds. The main action components are made of 4140 steel, while the stock unit is glass-reinforced nylon polymer. The twin magazine tubes are welded under the barrel.

These magazine tubes are loaded separately with a switch lever pivoting from one to the other. Each tube can be loaded as the shooter desires with any combination or mix of shotshells for hunting or defense applications. There is a pump mechanism action release lever forward of the trigger guard, which when pulled down allows the forearm to be pumped open like a traditional pump-action shotgun.

On top of the barrel is an integral 12-inch Picatinny rail for mounting electronic sights/optics, or BUIS flip-up open sights. A six-inch rail is available under the pump forearm for positioning a vertical grip and/or a laser device or flashlight. These design features make the KSG very easy to customize and accessorize by the end user.

The black matte finish appearance definitely projects foreboding to the weapon's application options. The factory is now offering an olive-drab-green version as well. The matte finish dispels any reflections to spook game or other targets.

I have had the Kel Tec KSG for about a month now and am still in the process of using it in the field and wringing out its shooting potentials. The KSG from the factory is pretty well set up to go to work as a stock firearm. With some practice in sighting and shooting, it can certainly be used as is without any major add-ons.

However, having said that, I tend to be an accessory nut, so I had to add a few things just to enhance the basic design and functional utility. First I added a set of the fine Magpul® BUIS flip-up open sights front and rear. These low profile sights are easy to fold down out of the way or push-button pop up when needed. The black versions I had in stock look quite handsome on the KSG and tightened down well on the rail.

Instead of the standard ubiquitous one-inch nylon sling supplied with the gun, I opted to install a Quake Industries soft touch sling to help offset the pull from the weight of the fully loaded shotgun. I will admit the attachment points for a sling at the KSG's muzzle are a tad bothersome for use and carry, but I see no other plausible option for this feature. Live with it.

I also played with mounting a Leupold electronic red-dot Prismatic sight on the top rail, but the jury is still out on that. My thinking is after all this is a shotgun with an open choke, so I expect the shot patterns to be somewhat widespread. Thus the utility of an electronic sight might be questioned.

If one were shooting buckshot or slugs or serving on night duty then I could see using such optics. During daylight hours, I would think most KSG shooters would use instinctive point shooting or a set of open sights. That's the shooter's option.

So far I have fired a selection of shotshells through the KSG. If you read other reviews of this shotgun, you will encounter commentary about how peculiar it is to load the twin tubes basically from the bottom, meaning the firearm has to be turned over to do this. It seems awkward at first, but practice improves the process.

The loading "port" does not have the feel of a traditional pump like a Remington 870. I found it difficult to load with the firearm upright. The good news is once fully loaded with 12-14 rounds, the shooter should be good to go for a while.

I noted that pushing shells into the magazine tubes was met with the sharp edge of something in the mechanism yet undetermined. I cut my finger the first time. I think there is a sharp edge on the lever that switches from one tube to the other. I plan to investigate this further and file if necessary. Also do not use brown cotton gloves, as they get hung up, too.

The KSG is a substantial firearm. Its weight when loaded has a welcome hefty feel to it. One might think a firearm of this short stature and design would kick like a blue tack mule, but it does not. If one gets a firm grip on this gun and leans into it, the recoil is not a serious issue. It points naturally and hits what it is aimed at.

Truthfully the trigger could be considered heavy, but personally I like that. I sure don't want a light trigger on a shotgun I am hunting with or using for other means. It is not a bad trigger, just a heavy pull. Again, practice will overcome this.

Now, to the one issue I have had thus far with the KSG. To date I have experienced a number of double-feed jams with this shotgun. I was using 2-¾ inch No. 6 birdshot, plastic hull shells for my initial shooting, because I wanted to test the shotgun for potential squirrel and rabbit hunting. It jammed over half the time.

Since then, I have consulted the owner's manual, which curiously fully describes the problem I am having. I hope to determine a fix. It seems that perhaps the KSG may have a propensity for the extractor claw to override the shell's rim upon firing and pumping the action. Thus, the fired shell is not pulled from the chamber as the loading ramp is trying to insert the next shell. This causes the jam.

The jam is not easy to remove. The owner's manual gives a full description on how to clear a jam, which incidentally I had already figured out sans the manual. Have a screwdriver handy to get the fresh shell pried out of the loading ramp.

Furthermore, I have gone back and given the chamber a thorough scrubbing with solvent and a brass 12-gauge brush. My hope is that this will smooth out the chamber, helping a fired shell to release better. I also have a feeling that with this being a new gun things will improve with more extended use.

Let me point out also that I reviewed several online videos on using the KSG and found that my shooting/pump-action technique might have also contributed to the jams. The shooter has to have a firm grip on this shotgun when firing and making the pump action extraction a smooth deliberate movement, not a jerking motion. A jerk of the pump forearm could cause the extractor to be too easily pulled over the chambered shell rim. I will practice my techniques further.

If you study anything about the Kel-Tec KSG, you are not likely to see any review relative to its use as a hunting shotgun. This may be primarily due to its configuration and also the cylinder bore choke. The ads for the gun typically show a para-military or security type application scenario. To me that is like saying that a combat Hummer could not be ridden around town on a Sunday afternoon.

I know enough now about the loads I have used in the KSG that it can definitely be used for hunting small game in trees or on the ground. With slugs and practice, this rig can take down a deer or a feral hog. Ditto with buckshot. Once the shooter learns where their bullpup prints its loads, then it is perfectly suited to hunting use.

I've said this before, but society has changed on public lands and private. Hunters need to be ready for anything, including personal protection if needs be. For this, the Kel-Tec KSG is the perfect tool. Go hunting with small game loads in one tube, and buckshot in the other. Switch back and forth as needed. Show me another smoothbore with that versatility.

The Kel-Tec KSG may take some getting used to, but with practice it might take a while to get that smile off your face.

PERSONAL PROTECTION HARDWARE: THE GOVERNOR IS ELECTED

For preppers trying to break free of the pending blitzkrieg or one already building into a mushroom cloud as the result of a recent SHTF event, busting out in a Bug Out escape mode can be a scary time. Everything is on the line as the threat increases and finally you give the nod to go. Now is the time. No more delays.

BASIC PREPPING ESSENTIALS: WEAPONS

The Bug Out location has long been chosen, prepared, and stocked. All the gear is packed, tote boxes are secured, the "land" rover is fueled, travel food and water is on board, along with maps, flashlights, and kids' books, and everything is checked and double-checked off the "preppers" essential have-to-have list.

The family is in tow, buckled in, the vehicle loaded to the hilt with every piece of necessary gear collected after countless hours of evaluating, shopping, and investing. Training courses complete, practice complete, family tuned in. You crank the engine.

With a whisper of a prayer for safety you ease the vehicle out of the driveway. The house is as secure as it ever will be as you glance back in your rearview mirror maybe for the last time...ever. A few neighbors out in their yards wave, timidly, cautiously, probably wondering why in the world you would be leaving in the middle of such upheaval. A couple people might even come out into the street to get you to slow down. You drive on, eyes directed ahead. There is no time for explanations now or goodbyes. You have bugged out.

Those were friends maybe, likely, you are leaving behind, neighbors, or at least acquaintances. Who knows when or if you might ever see them again? But what or who lies ahead down the road?

The unknown is a huge ingredient of terror. How do you protect yourself on the highway, or back roads, or at any needed stops along the way if necessary? And with what tool do you select for this personal and property protection task? Let me suggest a strong candidate to recommend to fulfill this critical role.

This is one unique handgun. It can be dispatched on many applications, but its top job is close quarters personal protection and varmint defense. The Governor is a traditional six-shot revolver with an elongated cylinder capable of handling .410 shotshells, the .45 Colt (Long Colt), as well as the .45 ACP. One of its marketing monikers is "mix six" meaning that any combination of rounds can be rotated in the cylinder at the selection of the user in tune with the intended applications for the firearm at any given time. The revolver can be fired in single or double-action mode. Cartridge rim clips are included for quick multiple round loads.

The "Gov" is a lightweight handgun tipping the scale at a mere 29.6 ounces. The frame is manufactured from a scandium alloy. The cylinder is made from PVD (physical vapor deposition) stainless, which is a thin protective coating on the metal surface. The barrel is 2-¾ inch with a Tritium Night Sight dovetailed into the muzzle end of the barrel. The rear "sight" is a fixed groove along the top of the frame. Again, this gun is intended for close encounters.

The grip is a soft recoil-absorbing synthetic rubber-like material and can also be ordered with a Crimson Trace Lasergrip red dot. The handgun's overall length is 8-½ inch, height is 5-½ inch, and a width of 1-¾ inches. The exterior finish is a handsome black matte.

Out of the lockable factory storage box the Governor has a great feel and balance. It is one of those types of handguns that just fits the hand, points well, isn't muzzle heavy or too bulky to tote. It will carry well in a variety of modes including a Bug Out Bag, bug out backpack, fanny pack, vehicle console compartment, side door panel, ATV handlebar satchel, shoulder holster, crossdraw, or hip carry on a good belt.

Since acquiring my Governor I also ordered a DeSantis Gunhide Dual Angle Hunter, right handed, thumb break leather holster. This model is custom-fitted just for the Governor and can be worn in the crossdraw position or on the hip with the gun butt forward. (www.desantisholster.com)

I prefer the crossdraw carry especially if sitting in a vehicle seat. In that configuration it remains hidden from somebody outside the vehicle, but easy to draw when necessary. This DeSantis holster is extreme quality made from premium leather with a soft suede lining assembled with tight, attractive wax thread stitching. This holster is both handsome and highly functional. I highly recommend it.

How should this Governor be sworn in? In travel on Bug Out Day the revolver should be in a place in the vehicle and near the

BASIC PREPPING ESSENTIALS: WEAPONS

user for a quick grab. That could be the driver or the passenger or both. Easily carried on the opposite shooting hand hip or located in a secure but discreet place in the vehicle, the Governor can be an effective deterrent once revealed, or an event stopping tool once deployed.

The critical decision for the user is to determine the "mix six" sequence of the three different rounds this gun is capable of handling. Likely any potential encounter with a source of serious trouble is going to be within 10 steps of the vehicle with the possibility of rapid advancement. Do you want to sting, ping, ding, or dispatch? Remember, this revolver can handle the .45 Colt, the .45 ACP, and the 2-½ inch .410 shotshell!

My own personal mix six combination when traveling is two .410 buckshot shotshells for the first two rounds backed up by four .45 ACPs. Since this is a revolver, you can load a wide variety of bullet types like hollow points or tactical loads without worrying about function as you might with a 1911 or other type semi-auto pistol for example. There are no safeties to forget or fumble, no magazines to engage, no slide to crank back and no fear of a boattailed bullet hang-up. With the SA or DA Governor you just cock the hammer and pull the trigger or go double-action. Just don't forget to roll the window down.

I believe in not reinventing the wheel whenever I can. There are plenty of, too many, ammunition and shotshell options when it comes to the .45 Colt or the .45 ACP and the .410 Shotshell. Winchester solves that decision-making when they packaged the PDX1 Defender Line of ammunition for personal defense with the Smith and Wesson Governor in mind.

Their .410/45 Colt Combo Pack puts together 10 rounds of PDX1 Defender 45 Colt, 225-grain ammo and 10 rounds of PDX1 Defender .410, 2-½ inch shotshells. You buy the ammo already packaged in one box ready to go. Just make up your own "mix six" load combination.

THE KIMBER 1911 10MM WHITETAIL "WONDERGUN" – PHASE ONE

Kimber pistols are the quintessential 1911. If you can get your hands on a Kimber factory catalog, for a pistol shootist it is like getting the old *Marshall Fields* Christmas catalog when I was a kid or even in my childhood hometown the Christmas flyer from the local *Western Auto*.

There are so many models in so many different optional configurations and calibers that it boggles the mind. This includes 13 categories of 1911 models, and 86 individual specific 1911 pistol models. Top that with sizes, calibers, finishes, and multiple feature options. That's a good thing.

Kimber currently is the No.1 supplier of 1911 pistols. That is in a marketplace with over 50 different manufacturers making the famed semi-auto handgun. Kimber owns over 40 percent of the total market, which is a tremendous achievement for any gun maker these days. I refer you back to the catalog or their website to see what I mean at www.kimberamerica.com.

I recently acquired a new-in-the-box Kimber Custom II-Stainless Target II in 10mm for an out-of-the-box project. I want to take a deer with it out of a ground blind over a lush green food plot where deer come in as close as 10 to 20 yards. I am not kidding, either.

We have over a dozen planted wildlife food plots with multiple types of stands available for deer hunting from ladder stands, tripods, and shooting houses. I also use a couple of different pop-up tent-type hunting ground blinds. Especially in the evenings, the deer file out of the woods and nearby overgrown habitats to take advantage of the green wheat, oats, clovers, and brassicas we plant. From a ground hide it is not unusual to have deer walk within yards of the blinds.

This makes for a perfect opportunity to take a deer with a handgun from a comfortable seated position using a Caldwell DeadShot FieldPod, a Primos Trigger Stick, or even conventional shooting X-sticks from inside the stand. You can actually sit back into the rear wall of the stand in the dark where deer cannot see you, yet you can see out both side and front windows. That is the plan anyway.

After considerable research for a semi-auto pistol to accomplish this plan, I pretty quickly narrowed my list down to just a few models. Two were Kimbers, and one was the older Colt Delta model, all chambered in the 10mm cartridge.

In a search at local gun dealers and after working three or four gun shows, the only 10mm semi-auto I could find was a Glock. I know this pistol is a good firearm, but it just does not fit my hand like a 1911 version. Also the Glock had fixed sights which I did not want for this task.

I discussed this mission with another friend of a friend gun dealer and they looked into ordering one. I was told 10 weeks minimum for the Kimber. That would have put my fall hunting plans out of the envelope of development time I needed. I continued the search.

And by the way, I have no clue why getting a specific Kimber model might be so difficult. It might have been more like a car dealer just wanting to sell what is on the car lot, not having to be bothered with ordering one.

I finally gave up a Saturday morning a few weeks ago to travel to another well-stocked gun shop some 40 miles away. To my surprise, the dealer had both Kimbers in stock, including two of the Stainless Target II models. I determined earlier from the Kimber factory catalog that the other model under consideration also had fixed sights. The Stainless Target II of course did come equipped with the excellent adjustable rear sight. That's the one I wanted for deer hunting.

The one Stainless Target II gun in the showcase had some handling scratches on it, and I am particular about buying a new gun thinking it should be pristine. The sales guy pulled the other one from stock and it was 100%. I did a little old-fashioned arm bending, drooling, begging and crying to get the retail price down just under $1000. That makes this Kimber one of the most expensive handguns I have ever purchased. That guilt and my high expectations were hoping for a positive outcome.

Specification-wise, the Custom II-Stainless Target II is pretty much a standard-sized copy of the original specs of the age-old Colt 1911 only modernized in many ways. The gun's height is 5.25 inches, length is 8.7 inches, width is 1.28 inches, weight is 38 ounces, with an eight-round factory magazine with an 18.5 pound recoil spring (which is why this model is a bit tough to hand cycle the slide).

The pistol sports a brushed satin silver stainless finish with black rubberized synthetic grip panels. It is a very attractive pistol. The five-inch barrel is ramped with a 16 left hand twist rate. The slide front is serrated for an easy grip. The trigger is an aluminum match grade version with three holes drilled for weight reduction factory set at Kimber's standard four to five pounds.

The black adjustable sights are Kimber's own. They are of heavy-duty construction with large adjustment screws for elevation and windage. There are easy to read arrows etched for the proper direction to rotate for adjustment. The rear sight has a square notch to match up with the front post. These are well-defined adjustable sights that are easy to sight downrange.

Some may ask appropriately about why I picked the 10mm for this project. Simple. In a 1911-type pistol configuration the 10mm is just about if not *the* most powerful cartridge available. It well outperforms the standards of 9mm, .38 Super, and .45ACP for purposes of hunting.

With factory ammunition loads of 170- to 180-grain bullets at just around 1,200 feet per second velocity, or the lighter 155-grain at 1350 fps, the 10mm is well suited for popping game-sized animals from coyotes on up to white-tailed deer at appropriate ranges. It is much more powerful than the .45ACP but is also considerably more flat in its trajectory, which is important for hunting ranges. In some circles the 10mm is compared to the .41 Magnum in terms of terminal energy ratings. That I am not sure about.

For my little project Hornady sent me two boxes of their excellent Custom handgun ammo in 10mm with their 180-grain XTP or *eXtreme Terminal Performance* bullets. Paper ballistics on this round include a muzzle velocity of 1180 fps with 1077 at 50 yards and 1004 at 100. Muzzle energy is set at 556-foot pounds with 463 at 50 yards. That would be my maximum practical range for shooting a deer out of my ground blind.

In prep for some range shooting, sight adjusting and such, I completely cleaned and lightly lubricated the pistol. At the range I initiated shooting the "out of the box" gun at 10 yards into a white pizza pie cardboard insert target upon which I drew black crosshairs. This was just to see where the pistol was hitting.

After an entire box of ammo expended with several sight adjustments mainly all to the right, I was finally getting into my standard of "shots inside a 10-inch circle" or the prospective kill zone of a white-tailed deer. I also set up a highway construction stand to shoot off of at 20 yards to repeat this level of acceptable accuracy.

I have to say, I do think this pistol as with most others I have "broken in" needs to be shot a lot more to be so-called loosened up. I plan to buy some standard hardball ammo just to run a hundred rounds or so through the gun before I go back with the hunting ammunition.

The trigger is very smooth with a great let-off. There are no problems there with me at all. The gun is very hard to cycle by hand and the slide release is extremely hard to release. I am used to 1911 slide releases letting go of the slide simply by pushing the release lever down. In this case I had to slightly ease the slide back from the muzzle to get the release to let go.

Thus far my only other issue with the Kimber and ammo combination was that several of the XTP bullets hung up in the ramp upon feeding from a fully loaded fresh magazine. I did not experience any hang ups when firing the pistol. Again, I think this situation will work itself out after more rounds are shot through the gun.

So, I am nearly at the end of Phase One of this project. I have acquired the appropriate handgun, quality hunting ammunition, have had a sight in session and have a ground blind set up to move to Phase Two. With any luck at all, I hope to prove out the results of all this in the coming days of the Mississippi white-tailed deer hunting season running through the end of January.

THE KIMBER CLAN OF 1911 PISTOLS

It seems most gun enthusiasts generally like all kinds of guns. Even so, we all have our favorites either defined by types of actions, calibers or gauges, or brand names. There are shotgun people, those for which "rifles rule," and the handgunners. Then you can further break down those general categories into more narrow pursuits.

I know smoothbore guys that think only side-by-sides exist. Others swear by over and under guns. Rifle shooters can be complete snobs for an action type, brand, or favorite hunting cartridge. Handgun people can go revolvers or pistols and then there are the crossover types that use them all. Each to their own certainly prevails.

Shotgun-wise I like a good classic double, but I love my Remington 11-87. Rifle-wise I am all across the board, loving Remington, Ruger, and Browning bolt actions. The 300 Winchester Short Magnum is my personal favorite white-tailed deer gun. I like the full daddy 300 magnum for elk. I love AR rifles, too, for deer, hogs, varmints, and paper.

I use, shoot, and hunt with all types of hand cannons. I like revolvers and pistols in a full range of calibers from rimfire to big centerfires. Years ago I started nursing an interest in 1911-type pistols, and now I guess have turned into a complete fanatic. In this field I have found a favorite and it's the Kimber line of 1911 pistols.

Kimber's history is an interesting study and for many a bit confusing, too. At one time, if I have the story right, guns were made under the Kimber name by two different entities, but now production has all been consolidated under one umbrella.

Kimber started out making high-grade hunting rifles over 30 years ago. As they diversified their product line they began manufacturing classic 1911 pistols. Today Kimber has become the world's largest producer of 1911 pistols and they lead the market share in this particular type of firearm. They currently produce 90 different models of 1911 pistols in 14 different factory-named categories.

That is quite an accomplishment given the explosion of interest in this pistol genre over the past decade. I am not sure what caused the sudden surge in demand for 1911s, but it has been strong for some time now. This buying energy of course encouraged numerous gun manufacturers, both large factory concerns and small custom shops, to produce a wide selection of 1911s in an endless array of standard and custom features.

All you have to do is go to a large regional gun show to see all the 1911 wares offered for sale. There are many, many high-quality 1911 pistols available on the market today from Colt, Smith and Wesson, SIG, Para Ordnance, Springfield Armory, Taurus, Wilson Combat, Iver Johnson, Ruger, Les Baer, Rock Island, Trident, Caspian, Dan Wesson, among many others I am not aware of, and then, of course, the Kimbers.

So, out of all those, why Kimber? I have owned, handled, shot, broken down and cleaned dozens of 1911s over the past 40 years, and I have seen many really good 1911s as well as a fair share of lousy ones. However, time and time again I continue to be particularly impressed with the fit, finish, detail to manufacturing, and overall quality of the pistols and the kit package that Kimbers are delivered in.

However, I am not so naïve as to believe there are not issues in delivering some 90 models of a particular design of a handgun. For example, just a few months ago I inquired of a retail gun dealer contact of mine about ordering a specific model of a Kimber, the Stainless Target II from the Custom II category of Kimber pistols. I wanted one in 10mm for another project.

After some talk with distributors, wholesalers, and even the factory, the dealer sent me an email to say that particular model (which by the way was a fairly common model) would not be available for 10 weeks. That was two and a half months by the factory estimation to deliver the gun. Perhaps it was the 10mm chambering that caused the lengthy delay. Lucky me, I drove 35 miles to another large retail gun store and they had two in stock.

This particular dealer in fact keeps one whole display case full of Kimbers, much to my delight. Even so, there are dozens of models of Kimbers listed in their catalog that I have never seen at this store or at guns shows in two states.

You have to obtain a Kimber factory catalog to really appreciate all the different types of diverse 1911 models they offer, even though they are all basically original 1911s at heart. For Kimber the models differ, too, by size. They manufacture the Ultra, Ultra +, Compact, Pro, and Custom pistol sizes.

Each size version has its own list of factory features and specifications as well, including grip length, slide/barrel length, caliber selections, fixed, adjustable, or night sights, grip materials, exterior finishes including mirror blue, black matte, stainless, Eclipse mirror finish, Desert Tan, plus two-tone finish models, weights (by manufacture material), left side or ambidextrous safeties, Crimson Trace lasergrips, and other proprietary design elements. Yeah, it is too much to choose from for sure, but enjoyable for 1911 enthusiasts.

Still, one wonders how a gun factory produces so many models with so many features and options and still keeps their quality control in check. While I have inspected nearly hundreds of Kimber pistols, I can say from outward appearance their new-in-the-box package delivers a beautifully made gun. However, I also know some of them don't always function to peak performance. Well, $100,000 Chevrolet Corvettes don't either.

Currently, my 10mm Stainless Target II is experiencing frequent stovepipe failures to feed from the magazine. I obtained two more stainless magazines and the issue persists. I am into trying a third brand/type of factory ammo and the gun is improving in function.

Several times now I have shot complete magazines of ammo without a stoppage. When it does hang up picking a new round out of the magazine, I merely have to nudge the rear of the slide to get it to close to lock down in full battery. Is this a bad pistol? Hardly.

With less than 100 rounds through the gun to date, I still think the Kimber needs more "breaking in." All the ammo I have found has a slight edge on the bullet that seems to hang up. I have yet to find any "ball" round nose 10mm ammo just to run through the gun. Next, I will try a complete cleaning and lubricating according to the factory owner's manual. This is a stainless steel gun, so lubrication can be a factor in reliable functioning. I am confident I will iron out the issues with this pistol.

The best news so far is that this is one accurate pistol. My goal is to take a close range deer with it out of a ground blind, and it certainly shoots plenty well enough to handle that task. The gun is a real pleasure to shoot. This is my first 10mm too that is touted for its snappy recoil. It handles just fine.

Incidentally, I have had similar issues with other brands of 1911s, so this certainly is not unique to Kimbers. Anybody that shoots the 1911 will understand that many of them require some tweaking and fine-tuning. At the very least one has to find the one or two ammo types that function best in each pistol.

For sure I am not a factory rep for Kimber, but I am sold on them. They have never sent me a factory loaner to use so I owe them nothing. If anything, they owe me. Hey, Kimber, send the hat and t-shirt!

If you like 1911s then do yourself a favor and go to your favorite gun outlet and ask to see one. I think you'll be as favorably impressed as I have been.

A COLLECTOR BEHEMOTH

The .41 Magnum is a bit of an anomaly. It was introduced in 1964, squeezed in between the .357 Magnum and the .44 Magnum. Some might have wondered then and even now what it was really created to do. Though never wildly popular, which is a shame, because the .41 is a grand balance between the other two powerhouses and is a sheer joy to shoot.

Remington Arms Company pressed for the full development of the .41 Magnum for large-framed revolvers akin to the ubiquitous Smith and Wesson Model 29 made famous by Clint Eastwood as Dirty Harry. This exceptionally well-made large handgun became the gold standard for magnum revolvers "back in the day" and that status remains today.

The Smith and Wesson "N" frame double-action revolver was chosen for the .41 Magnum originally in 1964 in the Model 57 DA Revolver, target version, blued with 4-6-8-3/8-inch barrels, adjustable sights and large target-style wood grips delivered in a blue velvet-lined wooden case. It was a slick set up for sure.

Later that year Smith and Wesson brought out the Model 58 Military and Police DA Revolver without adjustable sights having simple fixed sights in 4-inch barrels only in a blued finish with checkered walnut grips for law enforcement use. The original model remained in production only until 1982. After that these became instant collector's items and still are.

The Model 58 is a gun I sought after for well over 25 years and finally connected on one in 2005. Prior to that, I had never seen but one after attending well over 200 gun shows. As it turned out one day I was on my hands and knees looking into the back of a glass gun case at a local gun shop when I spotted what I thought might be a Model 58.

When I asked the shop owner about it, his immediate reply was that "I sure had an eye for guns." The used 58 was a bit worn. Whoever had owned the one I found installed a red insert into the front sight. I would rather have the original sight but have been unable to find the part.

The handgun was rather crudely stamped "S.A.P.D". I learned later after some research that this was the San Antonio Police

Department, one of the few entire police departments in the country that went to the Model 58, .41 Magnum. It was likely one of the last, too.

Alas, it proved too difficult for the average street cop to shoot well. Those big 210-grain bullets also had a tendency to sail thorough obstacles and keep on going. The Model 58 did not last long as a law enforcement choice. Most LE agencies fell back to the .357 Magnum or else moved on to the new semi-auto pistols chambered for the 9mm and a new era in law enforcement firearms was born.

I have shot my Model 58 several times and it reminds me of a mini flamethrower. This baby definitely has a kick on both ends. Even with just a 4-inch barrel though, it is remarkably accurate for short range work that might have typically been used for police work or overkill for self-defense. It is authoritative.

The .41 Magnum uses a standard 210-grain bullet, either lead round nosed or more likely a jacketed soft point or jacketed hollow point. It generates roughly 1,300 feet per second muzzle velocity and 1,062 FPS at 50 yards. The muzzle energy is rated at 788 foot pounds, with 630 foot pounds at 50 yards and 526 foot pounds at 100 yards.

This made the .41 a perfect fit for mid-range game hunting for deer-sized game as well as an ideal round for pig hunting. Using heavy cast or jacketed soft points, the .41 can reduce a sizeable wild hog into a Sunday bar-b-que quickly with a well-placed shot. The .41 is also suitable for popping coyotes, fox, bobcats, and other such varmints for handgun hunters looking for a real challenge.

Today, there are not many handguns produced in .41 Magnum. Ruger still makes a single-action Blackhawk for it in a couple of different barrel length, the 6-½ inch being the better choice. Smith and Wessons can still be found, but the original early Model 57s are very hard to locate. I saw one with a 4-inch barrel virtually new in the box a couple of years ago, but the $1,500 price tag chased me away. For now I guess I'll have to be content with my old Model 58 flamethrower.

THE GLOCK GENERATIONS

After several years of working with the Glock Pro guys at gun shows installing sights, triggers and accessories on Glock pistols, I grew to have an appreciation for the Austrian pistol. Personally, I still do not like them, but there is no denying the success of the pistol design and its widespread use. Glock is still probably the most chosen handgun for law enforcement for many reasons.

Even so, the design is not perfect and even Glock recognizes that features and specifications need to be changed or upgraded from time to time. Mostly that is driven by their customer base of military contracts first, then law enforcement contracts second. The public side, commercial use of Glock pistols has never really been a high priority by the company. Imagine that.

Hence the reason for now four different "generations" of Glock pistols, generations being defined by the company as model specification changes, design alternations, upgrades or such.

The present iteration of the Glock is the "Gen 4" model. But, ironically, the factory is still turning out Gen-3 versions and this generation of Glock pistols is still selling roughly one-third of the total production. Why is that?

First, look at the changes they made to the Gen-3 to make the Gen-4. The most noticeable change came with the slight downside of the frame and the smaller grip. This was to accommodate law enforcement personnel that found the Gen-3 harder to hold onto. Added to this change came the interchangeable backstraps that could help fit more shooters. The grip surface stippling was also made more aggressive with a tiny field of gripping dots.

The "soft" plastic frame of the Gen-3 was made harder so the Gen-4 could accommodate tactical lights without the frame flexing too much, especially with the .40 S&W models. The recoil springs were also altered to handle the .40. Ironic now given the fact that many LE outfits are trading in the .40s for 9mm guns.

BASIC PREPPING ESSENTIALS: WEAPONS

The magazine release button was made larger. Some like this, but many Glock shooters do not. They prefer the Gen-3 magazine release which is more extended.

Some Glock shooters have some aspects of the Gen-3 that they wish had been improved on the Gen-4 but were not. Tops on this list is the trigger pull and the sights. Maybe these items will be attended to when or if a Gen-5 model ever comes out.

I WANT TO BE A .300 BLACKOUT WHEN I GROW UP

Survivalists, preppers, and bug out artists are all about the .223/5.56 in an AR-15 platform. Though this is a hard combination to argue against or to not have at least one set up this way, there are other viable options. Some of these options are decidedly more effective at what they do best.

First, let me pitch the value of a .30 caliber bullet over the .22 caliber. Remember even the .223 and 5.56 are nothing but .22 caliber bullets. Typically the bullet weights are decidedly light, too, usually in the 45- to 60-grain categories. Driven at high velocities they can do funny things on hard surface targets especially at longer ranges, say well beyond 200 yards.

Now, don't get me wrong. I do not want to get shot with a 5.56 round at any range at any time. However, I might want to take my chances especially if I am behind cover rather than be popped at with a .30 caliber pill weighing twice as much.

The .300 AAC, or Blackout as the round is referred to interchangeably, was created originally from the wildcat round known as the .300 Whisper. The .300 AAC Blackout is also technically referred to as the 7.62x35mm so it can be sized up to the Soviet Block arms and other developments around the world. The AAC comes from the cartridge developer Advanced Armament Corporation.

This new .30 cal round was developed initially as a consideration for the replacement of the military 5.56. In the final analysis the total cost of switching out all the AR upper units as well as all the support gear, carry webbing, and whatnot, just proved too expensive for a downsizing U.S. military establishment. Thus, the .300 Blackout was left to the commercial market.

A standard load for the .300 BLK uses a 125-grain bullet driven at 2,215 feet per second, generating a terminal energy rating of 1,360 foot pounds. The nice part is that more new loadings are coming out all the time that are much more economical. Also the .300 Blackout uses the exact same AR magazine as the 5.56.

This .300 provides all usable capabilities in a shorter, more compact, lightweight, durable and low recoiling package. This makes the .300 Blackout a super alternative SHTF round to the .223/5.56.

THE "NEW" INLAND M1 CARBINE

My dad was famous for saying "That looks like a solution to a non-existent problem." What he meant by that was there seemed to be a number of products developed over the years that were supposed to fill a void or need that did not really exist. In many of these cases, it was not too awful long before these items disappeared from the marketplace altogether.

With the introduction of a new version of the old version of the M-1 Carbine, short rifle in .30 Carbine, I wonder if this weapon falls into the category of producing or bringing back something that really isn't needed.

By reading the history and performance of this unique military carbine, light rifle, I wonder from its origin if it was really needed in the first place. I suppose so. It was created to fill a sort of void for the military somewhere between the sidearm of the day, the 1911 Colt, .45 Auto and the M-1 Garand battle rifle in 30-06.

Officers, pilots, paratroopers, and others not consistently on the front battle lines wanted a handy, lightweight, short rifle to offer more ballistics firepower than a sidearm, but not a heavy, cumbersome weapon like the M-1 Garand or the BAR. Thus the M-1 Carbine was developed.

Without a doubt this rifle served some purpose for the military, but reading in depth on its use, it really did not get glowing reports of effectiveness. In fact, during the Korean War the rather paleface .30 Carbine round was readily found to fail to penetrate the heavy quilted coats worn by the Chi-Com troops pouring over the Parallel.

The M-1 Carbine still sees service around the world in some minor military roles or some law enforcement units, but it has long been retired by the U.S. Army. Ironically, though the rifle still enjoys good interest among collectors, it never really found favor with post-military troops or others for either recreational shooting or hunting.

Recently I read a report evaluation on the new Inland Manufacturing version of the M-1 Carbine, and the report was not very flattering. The metal work was found to be extremely rough and it failed to function during repeated range tests. The stained oil finish on the stock came off on the hands and cleaning rags. The magazines did not fit well or function reliably.

So, why produce such a firearm for the public? Consumers will be the judge of that. It may just be another solution to a non-existent problem.

IVER JOHNSON'S LONG SLIDE 1911

So, what does the world need with another 1911 design handgun? Well, duh, that is like asking why we need another anything. Sure, we can get along fine without another chicken recipe, tactical knife, survival book, hiking boots, and more assorted gismos and gadgets. But, heck, this new stuff not only drives our interest as preppers and survivalists, but it generates a huge engine for our economy. We hope that engine runs well as long as possible.

A. FINALLY AN AFFORDABLE LONG SLIDE

I have never had the good fortune to work with a long slide 1911 until now. I have always been enamored with the longer six-inch 1911, but these guns have always been custom shop versions from somebody like Les Baer or Nighthawk.

Unfortunately the prices of these specialized custom pistols rival the sticker price I paid for my first new car, a 1969 Pontiac GTO for $3,600. These may be exceptionally well-made pistols and perform like a Swiss watch, but most of us preppers or survivalists simply cannot justify paying that much for a personal firearm. Well, now there is another option.

B. IVER JOHNSON BACK ON-LINE

Iver Johnson Arms and Cycle Works began tinkering with the idea of making guns as far back as 1871, but began manufacture of firearms in earnest in 1883. That organization ceased operation in 1993, but by 2004 was back making guns again. Over this tenure, the company made all types of inexpensive revolvers and some models of pistols as well as other guns.

Today's Iver Johnson Arms, Inc. (www.iverjohnsonarms.com) makes their new line of 1911 pistols in the Philippines. Incidentally, in doing some background research, I discovered that more 1911s are made there than anywhere else in the world just in terms of production numbers. I did find that rather surprising.

So, knowing that, I would be lying if I did not wonder what the quality of such guns was, coming from that location not particularly noted for its manufacturing prowess. But, despite that, I wanted to try a six-inch slide 1911 and Iver Johnson was producing one. So, I had my dealer order it, despite a tinge of nervousness.

When the pistol arrived, the dealer called. I asked my gun bud what his initial impression was of the Iver Johnson 1911. He was

BASIC PREPPING ESSENTIALS: WEAPONS

delighted in what he saw. He reported the out-of-the-box fit and finish was excellent. He bragged on the finish and the adjustable sights. I felt a sigh of relief for sure. The unknown is always a bit scary.

C. IVER'S EAGLE XL

Iver Johnson Arms' new 1911 long slide has been titled the *Eagle XL*. It is the full six-inch barrel and matching slide in a matte blue finish with very attractive walnut grips. The slide bears a fully adjustable rear sight with a dovetail front sight. The slide carries both front and rear angled serrations that are easy to grip to cycle the slide.

The frame comes with an extended slide stop and thumb safety. There is an aggressive Beavertail grip safety with a memory bump. The trigger is a modern three-hole style with a combat-type skeletonized hammer. The ejection port has been lowered and flared.

What all these included factory features say to me, is that Iver Johnson has made a real dedicated effort to produce a more or less standard pistol package with a number of custom features. Had they added ambidextrous thumb safeties and included two factory magazines the pistol would really be special. This base model retails for $845, but that is list price. Shop around.

Though I am just giddy about the six-inch slide, it does serve several practical purposes. First is the increased sight plain. If your eyes are aging or you wear glasses, that extra inch can help sight focus. The extra-long slide also adds weight up front to assist with muzzle control flip. The pistol's overall handling characteristics are improved as well. It holds well and points better. That is the first thing I noted. All in all, the Eagle XL is a finely made pistol with lots of features. But does it shoot?

D. RANGE BANGS

The mag for the Eagle XL holds the standard eight rounds. They hand load easily. The magazine locks up firmly with a secure snap. The slide charging is typical 1911, but the serrations help, especially those lines cut up front. The Eagle XL has one of the easiest slide cycles I have ever experienced. It was smooth, even, and easy to manipulate.

I consider this pistol good for personal protection, but borderline for on-body CCW. Beside the car seat, in a crossdraw holster on an ATV or horse, or in a slip case at the ready, the Eagle would be great. It would be an imposing defensive weapon upon perpetrator view.

For the initial range break in testing, I used simple, standard ball .45 ACP loads with 230-grain bullets, Remington UMC and Blazer. I like to see if a gun will work with basic ammo before I move on to the max expensive specialized self-defense loads.

If a 1911 will throw ball ammo into a 10-inch pie plate at defensive ranges then I can get down to business with a hollow point or defense load. By the way, recent self-defense incident shooting data has reported that most one-one-one shooting encounters occur at three to seven *feet*. Yep, feet. I usually shoot such a pistol at 10 feet to 10 yards. I may have to modify that approach.

During my range session, I used the Eagle XL magazine supplied with the pistol, but also two other magazines just to test their accommodation. One of the extra mags I used was a Colt and the other was a totally non-descript 1911 magazine. All three magazines locked up, fed without issue, and locked the slide back open at the last shot. That is a good sign.

The Eagle shot all rounds without issue, everything cycled fine, and there were no hang ups, stovepipes, or stoppages. Trigger pull was for me ideal, smooth with no grind or glitches. Shooting at a defensive 10-feet, the accuracy was exceptional in my opinion. This is one fine-shooting 1911. It is definitely a keeper and the six-inch slide is just a welcomed bonus. Check it out on-line, then get your dealer to order one.

CHAPTER 2
PREPPER AMMO AND CARTRIDGES

A SHORT LIST OF SURVIVAL CARTRIDGES YOU MUST HAVE

There are far more firearm cartridges available than we will ever need. Just flip through the pages of a copy of the renowned book *Cartridges of the World* by Frank Barnes. You will enjoy reading details and specifications on over 1,500 individual cartridges, and more have come out since the latest edition of Barnes' book

Even within this exhaustive volume Barnes lists current American rifle cartridges, obsolete American rifle cartridges, wildcat cartridges, proprietary cartridges, handgun cartridges, military rifle cartridges, and several other CHAPTERs covering a wide spectrum of foreign cartridges as well. There is plenty enough of them to satisfy any shooter.

But truthfully, preppers and survivalists don't need near that many cartridges to handle a multitude of SHTF scenarios. Accordingly though, I fabricated this list of highly recommended cartridges that every prep planner should strongly consider adopting. My suggestions are based on over 50 years of shooting and 35 years of writing about guns, cartridges, hunting, personal defense, and gun collecting.

That does not make me an expert in the field, however, and having even ventured forth in writing this CHAPTER, I do so knowing full well that it will be highly critiqued and criticized. Everybody has their favorite guns and cartridges and my list will not encompass them all by any means. So, consider my recommendations and look at others, and then make your choices. Trust me though, this minimal list will cover all your necessary bases.

For plinking, fun shooting, potting for the skillet or bar-b-que, you simply have to have a rimfire rifle and ideally a rimfire handgun or two. Pick the long rifle rimfire cartridge and don't bother with the others.

Rimfire guns can be used for an endless number of roles for a survivalist. For teaching, learning, passing on safe gun handling skills, shooting practice and minimal self-defense protection, a good rimfire rifle or handgun is essential. Teach the wife and kids to use them safely and effectively as well.

Use them for hunting, taking out varmint intruders near camp or around the Bug In residence, and for potentially chasing off other bothersome uninvited guests, trespassers, or poachers. Any shot fired is better than no shot fired when appropriate.

Up until the past year, rimfire ammo was cheap. There is *no* cheap ammo of any kind any more. Still, rimfires are cheaper to use than centerfire guns, so plan on having a good quantity of ammo and several rimfire guns in stock.

Many will argue the 9mm is too light and ineffective. I keep reminding myself that the U.S. military chose the 9mm for universal use. That is an endorsement I find difficult to ignore. Numerous law enforcement and other government agencies use it, too. And a double stack magazine can sure put a lot of firepower in your hands.

The standard 9mm load uses a 115- to 124-grain load pushed to 1,225 MV. There are so many 9mm ammo loads from the factory available, I cannot list them all here. Winchester's ammo catalog alone lists 17 different loadings for the 9mm in bullet weights from 105 grains to 147. Additionally there are all types of bullets available and many new self-defense loads as well. The 9mm is highly respected and trusted.

One of the main reasons to use the 9mm for prep plans is the sheer fact of its universality in that the ammo is widely available and fairly affordable. I have never been in any ammo store anywhere or any ammo website that did not have multiple selections of ammo for the 9mm. During a SHTF grab, 9mm will be the first to go.

This is my personal favorite centerfire cartridge in a semi-auto pistol. The 1911 Colt and clones are my own choice to deliver these heavy 230-grain pills. I found the .45 ACP just as manageable to shoot as the 9mm with less snap at the wrist and less hypervelocity muzzle report. Any intruder or undesirable chest-centered with one of these will certainly alter their thinking about forward aggression.

The .45 ACP ammo-wise is nearly as prolific as the 9mm. Winchester has 12 offering and other manufacturers add greatly to that list. Usual bullet load in the .45 ACP is the 230-grain Full Metal Jacket or "Ball" ammo. There are hollow points, Silvertips, flat nose, jacketed flat points and other bullet choices in weights from 170-185-230 grains. As with the 9mm, ammo makers have brought out a bunch of choices in personal protection rounds and self-defense loads.

The .45 ACP is a top choice for a heavy-duty close quarters personal defense and protection cartridge. Though there are a few revolvers that can handle it, it is best launched in a semi-auto pistol package. I will not deny that the .45 ACP takes some practice to use well, but then to my way of thinking all firearms do.

The first SHTF prep rifle cartridge of choice is the .223-5.56. Say what you want but the .223 is widely available in dozens of AR-15 rifle platforms in all price ranges and customized accessory combinations. The .223 can also be mated to a classic bolt action rifle and large scope for a rifle package perfect for targeting longer range targets.

The usual .223-5.56 cartridge load handles the standard 55-grain bullets. There are also 35, 40, 45, 50, 60, 62, 64, 69 and 77-grain

BASIC PREPPING ESSENTIALS: WEAPONS

bullets in many styles. The full metal jacket bullet is the standard but loads can be found in hollow points, boat tails, match, green tip steel penetrator, flat base tip, plus some hunting loads like the Sierra Blitz-King, Barnes TSX, and many other choices. Muzzle velocities range from 2,500 to 3,600 feet per second.

Because I personally lean toward the AR-15 platform, I think the .223-5.56 is the ideal mating for a great SHTF rifle set-up. Add a red dot, conventional scope, or an electronic optic like an EOTech and you have a perfect rifle for property protection, defense against uninvited two-footers, or others.

The selection of both rifles and ammo here are endless. Literally dozens of accessories are marketed to make up your AR just the way you want it. And that is half the fun of this package.

Also known at the range bench as the 30-06 mini, the .308 can do anything its bigger brother can do, but is also more accurate. Every SHTF plan needs a heavy-duty centerfire rifle in either a semi-auto MSR (modern sporting rifle) and/or a bolt action sniper type/hunting rifle rig. The .308 is the "king of takedowns" and should be considered for serious defense work and a serious hunting round for supplying fresh meat to the Bug Out camp or other alternative SHTF living arrangements.

Weapon-wise there are now several great MSRs in .308 as well as some excellent bolt action hunting rifles and combat-type bolt rifles on the market. Look at Ruger, Springfield Armory, Rock River, and others. Exceptional bolt actions can be found at any gun shop, gun show, or firearms outlet.

The standard .308 Winchester utilizes the basic 150- or 180-grain bullets. The tradeoff of course is that the lighter bullets travel faster and shoot somewhat flatter. The heavier bullets produce greater terminal energy. Factory loads for the 308 are simply too numerous to list here. Check into ammunition by Hornady, Remington, Winchester, Federal, and others.

Mate a quality .308 rifle with an equally quality riflescope like a Leupold, Nightforce, Zeiss, Nikon, Bushnell, or similar high-end optics and you will have the foundation for an exceptional rifle rig for serious heavy gauge work.

You will need a shotgun and the 12-gauge is it. Find yourself a high-quality pump action, or semi-auto action shotgun that is well built and has a reputation for reliability and easy maintenance. All of the name brand manufacturers have at least one model.

A good shotgun with a standard length barrel of 28-inches that uses screw-in chokes can be used for bird and small game hunting, bigger game or deer hunting with a barrel suitable for buckshot, or slug barrel for big game hunting. Shotguns are exceptionally good for close quarters self-defense, defense of property, and for handling cocaine-addled zombies banging at the front door.

There is no reason not to have a shotgun for which the barrels can be changed out to suit the immediate job at hand. The Remington 870 is perfect for this as extra barrels of all types are readily available in the marketplace. If you'll stick with barrels that use a change-out choke system you add further flexibility to your various missions.

Stock up on several kinds of 12-gauge shotshells. Get light loads for small game and bird hunting. Get shells for waterfowl and turkey. Buy additional loads using buckshot and slugs. Keep on hand loads for self-defense, guarding, and protection. These should cover all your shotgunning essentials.

Well, there is my list of the essential SHTF cartridges you might want to consider compiling for a wide variety of survival tasks. Are there others to consider? For sure you can add to the list for consideration the .380 ACP, 40 Smith and Wesson, .38 Special, .357 Magnum, .357 SIG, .41 Magnum, 10mm auto, .45 Colt, the 7.62x39 (AR-47 round), 300 AAC-Blackout, 6.8 SPC and basically any "deer" rifle including the 30-30 if that is all you have or can afford.

Certainly you will want some kind of self-defense, personal and property protection arsenal as part of your prepping plan. You can go basic or you can go overboard depending on your budget. Just keep in mind the cost factors each time you add another cartridge to the mix. You will be talking extra ammo, magazines, cleaning supplies, holsters, and everything else that goes along with practice to keep proficient and prepared. Consider the list offered here as a solid starting place or the end-all you will need.

ASSESSING NEEDS TO BUY AMMO

How much ammo do you need to buy? At what cost can you afford to add to your stores? First, consider what type of shooter are you and at what level your shooting demands are? It might make a huge difference if you are a simple deer hunter or a high volume one such as a varmint hunter, or a target shooter, competitive shooter, or a survival prepper.

If your primary concern is for self-defense at home or on travels, or you have to conduct security surveillance on properties, rural lands, Bug Out camp or other circumstances, then your needs may be for more highly specialized ammunitions which often translates to being more expensive. Some balance has to be struck between needs and wants when it comes to spending money on ammo.

With all of the scares these days from manufacturing capacity fluctuations, importation complications, the political climate, military supply demands, reports of huge government agency ammo stockpiling, wholesalers holding back, price fixing, dealers restricting sales amounts, federal agency threats to end certain kinds of ammo production, and all else, it is no wonder ammo consumers are in a panic mode to stock up.

Initially all shooters need to take a complete inventory of the ammo stores they actually have on hand. If you reload, then do the same for powders, brass, primers, and bullets. Know which calibers you shoot most and also the ones used the least. With this information in hand one can better make decisions on what to buy and how much of it.

Can you determine how many rounds of ammo you use in a year under normal circumstances? For example, I deer hunt with either the .300 Winchester Short Magnum (WSM) or the .35 Whelen. In a typical year of deer hunting I will fire less than five rounds of each cartridge. So, I am not pressed to have a big supply of this ammo on hand.

Survival preppers on the other hand work to maintain a sizeable stash of ammo for those just-in-case SHTF events. Take your own pro or con approach to that line of thinking but the ammo demands are real. Many preppers try to keep a minimum of 5,000 rounds of each of the primary cartridges they use. This might include the .22 rimfire, 12-gauge shotgun, .223, 9mm, 40 S&W, .45 ACP or others. This can make for a considerable cache of ammo. I know some preppers that keep five times that amount of ammo.

So, set your priorities on what to stock, at what level, and work on a realistic budget to acquire what you need or desire to keep on hand. Then you can proceed with your shopping strategies to build up your inventory to the level you want.

Before you set about shopping for ammo, decide what brands, types, bullets, loads, and other specifications you prefer to buy. You may need to standardize a few things. There are four major domestic brands of ammo including Winchester, Remington, Federal, and Hornady. Even among these brands there are multiple offerings of bullet types, weights, and specialty ammo for hunting of all kinds of levels of game animals, self-defense ammo, target shooting ammo, and cheaper plinking stuff. Figure out first what exact load you shoot before you buy anything in quantity, though you may always want to test out some new loads by buying a box or two.

There are many other ammo brands as well, a sort of second tier as it were to include brands like PMC, Speer, DRT, Magtech, Armscor, American Eagle, PNW Arms, and Buffalo Bore. This is not lesser quality ammunition, just not as well known in the overall consumer marketplace. You may never find these on a local dealer's shelves.

Furthermore, there are several foreign brands of ammo that do not rate as well. Basically if a foreign ammo uses steel casings, Berdan primers, corrosive powders, and odd weights of bullets, then be careful trusting it as your primary ammo. Modern guns were not really intended to fire steel casings with just a lacquer coating on it.

Most current production chambers were designed to function with brass cases that flex and seal fit to the chambers upon firing. Steel cases can score and ruin the chamber of a gun. If you shoot an old AK-47, SKS, or something, then use steel casing ammo if you want, but reserve for your best firearms brass-cased ammo only.

Shopping for ammunition these days is made a lot easier and sometimes cheaper with the availability of the many Internet sources that sell shooting supplies. The trick is to monitor these sites all the time to check what is in stock, what is the price, is it ever on sale, and what shipping restrictions or costs are involved. This may require signing up for email notices of supplier newsletters, sales flyers, and promotions of all kinds. It takes time to assess these, and compare various sources against each other.

It is always advisable to take advantage of any ammo sales at local retailers or big box stores. When they send out a sales flyer with ammo marked down, expect a "run" on the counter and stocks to go fast. I live 20 miles from a Bass Pro and by the time I can get there the sale ammo is almost nearly always already sold out. Good deals go fast.

Gun shows usually offer a good selection of ammunition at fairly competitive pricing, but rarely do I see it marked down per se. You just have to shop gun show tables very carefully and be prepared to buy if the deal is particularly good. If you hesitate, someone else will surely take advantage of it.

Though it is prudent to pick up a box or several of often-used ammo as needed when you find it, it is better to buy high volume use ammo in quantity. Say, for example, if you shoot a lot of .223/5.56, then you can often find very desirable pricing if you purchase it by the 1,000-round case lot. Again, shop around.

When you shop Internet ammo sites, be careful of the shipping charges. Ammo is heavy and very often the shipping charges are too. If you watch carefully there are some Internet suppliers that occasionally offer free shipping. Strike while the iron is hot, or lead in this case. Be sure you take note of the shipping carrier, too. If it happens to be Fed Ex, you might have to be home to sign for it. If you work, that can be a problem.

I do notice that most Internet ammo outlets now ship the commercial product brand packaging inside of another box with protective wrapping. Though a delivery driver has to guess what is in the box, it is not made obvious. I am always a little nervous with them leaving that box by the front door, but so far, my neighborhood has proven safe for this.

So, don't be perplexed or deterred about buying the ammunition you need. Prices go up and down all the time based on the paranoia climate of the country. Stock some of what you have to have on hand, then be patient to buy more when conditions are favorable. Be sure to store all your ammo securely, safely, and in a climate-controlled area. As I always tell my buddies, "If you are worried about the state of affairs in the world, just buy more ammo."

IS THE .380 THE NEW 9MM?

Yeah, this is a serious question. No fooling around. As a long time outdoors/gun writer, I watch all the trends in the firearms industry. This includes guns and ammo popularity reflected in national sales, new gun models coming on the market, as well as new ammunition.

The last decade or more has shown tremendous energy in creativity and productivity when it comes to the firearms industry. Sometimes it moves so fast it is hard to keep track of everything happening. It's not only challenging, but fun to read all the material coming out, tracking the gun phenomenons across the country, and then letting my readership in on a few insights I have noted.

BASIC PREPPING ESSENTIALS: WEAPONS

One of the trends I have been monitoring over the past couple of years is the number of new handguns that have come out chambered for the .380 ACP. Most of these new guns have been classified or labeled as "pocket pistols" or CCW-mode, meaning for concealed carry. That is the marketing ploy angle anyway, and frankly it works well for pistols chambered for the .380.

Now if you regularly read any gun magazines off the rack and study what is going on, you know too what is happening in the .380 realm. There are literally dozens of small pocket pistols available all of sudden in .380 ACP.

Of course, the .380 has been around forever, since WWII in the Walther PPK and later in the PPKS. The .380 was also initially referred to as the 9mm Short. This is not the place for a history lesson, but you might want to look up all that on your own.

Just in the last couple of years the .380 market has become flooded with some excellent pistols. Names like Sig-Sauer, Beretta, Accu-Tec, Diamondback, Ruger, Smith and Wesson, Colt, Kel-Tec, Glock, Bersa, Taurus, and Kahr come to mind immediately. Undoubtedly there are others out there that I can't recall, forgot about, or don't know about. Of course, also note that all of these handguns are semi-automatics, hence the classification as pistols. Pistols are red hot right now. I suspect that will continue on, too.

So where does all this rather sudden interest in .380 pocket pistols come from anyway? First and foremost I think the main reason is for self-protection and protection for the home. While the .380 might be thought of as a "lady's gun" it has much more application than that given the development of new defensive ammo choices in the .380. I suspect a fair share of law enforcement officers are using .380s for backups as well.

I also think the .380 is a decent choice for outdoors enthusiasts to carry when hunting, fishing, and hiking or such if they are out in public areas. These days you simply can't be too careful going anywhere. Put one in your gear bag or in the vehicle when you travel. The .380 is a reasonable compromise for these uses.

So, as in the title, I pose the question given current market trends. Is the .380 the new 9mm? If I imply a replacement, then I doubt this is the case. It is a practical matter that most, but not all 9mm pistols are larger, heavier, and thus slightly more difficult to wield.

Not so if you just practice with them, but they recoil more, are louder, and are more difficult to conceal easily. The 9mm is hardly going away. I will say though that a 9mm in some of the new micro-pistol packages can be quite a handful. They are close quarters pistols though and like the .380 are meant for short range engagements.

However, hard to argue that a neat little .380 can fit right into your pants pocket, a purse, a waistband holster, book bag, backpack, fanny pack or such. They are small, light, and easy to conceal. With practice and proper safety instruction, they are easy and pleasurable to shoot. The trick is getting used to the controls as second nature.

Reality is, too, that the .380 is no 9mm in terms of terminal ballistics, but it was not meant to be. Remember the .380 is the short version of the 9mm, implying lesser power.

With new ammo revelations recently though, the .380 has gained some respectable ground in terms of defensive power.

In general terms, picking a universal-type load each for the .380 and the 9mm for comparison, how do they rank? Many .380 loads use a 90-grain bullet with a muzzle velocity of around 1,000 fps. Its muzzle energy is listed at 200-foot pounds. The 9mm's "standard" load is a 115-grain bullet with an MV of 1,160 fps and a ME of 345 ft. lbs.

You make the judgment which you would rather have. There have been many complaints over the years about the ineffectiveness of the 9mm. My own opinion is that the shooters expected too much from it or overextended its logical capabilities. Again, both the 9mm and for sure the .380 are short range affairs.

New ammunition developments have greatly improved upon the numbers listed above. So for the .380 check out Winchester's SXZ, Federal Guard Dog, Barnes Personal Defense, Hornady Critical Defense, and other new loads for the .380 and the 9mm.

This new "defensive" ammo genre really ramps up the effectiveness of personal defense handguns including the .380s. It only makes sense to use the most effective ammo you can buy if the setup is going to defend your life and protect family and home. You are going to want one of these along on all outdoor adventures.

So, is the .380 the new 9mm? Well, I have to hedge the bet by saying yes and no. It rather depends mostly on the application. For an outdoors person toting a backpack or fanny pack, a small concealable .380 is a good choice. It could be used to dispatch wounded game or offer defense against an aggressive trespasser, hoping no incident would ever come to that.

Putting it all in perspective, the .380 is better than a .22 rimfire, or something like the .25 or .32 ACP. But it is not as powerful as the 9mm, the .38 Special, .357 magnum, or some other larger caliber choices. Still, in a small concealable pocket pistol package, the .380 sure does offer a lot of positive compromise.

TRIALS WITH THE .35 WHELEN

When the Department of Wildlife, Fisheries, and Parks permitted the use of centerfire, single shot rifles for the primitive weapons deer season, the whole face of deer hunting changed in Mississippi.

The analysis of that move proved two things. It killed the sales of black powder guns. It greatly increased the interest in hunting the primitive seasons. To be honest, gun dealers sold more of the new "primitive" weapons than they ever sold in black powder rifles.

Initially the minimum for the new primitive rifles had to be at least .38 caliber, but there were only a few rifles that met all the criterion. A couple years later the rule was adjusted to allow rifles in .35 caliber. This in effect added only one new choice: the .35 Whelen, which was the only .35 available in the approved primitive weapons configurations. But, boy howdy, what a choice that was.

The short course on the .35 Whelen is that it is simply a 30-06 necked up to .35 caliber. Recent research confirms that arms tinkerer Colonel Townsend Whelen was instrumental in development of the round.

It started of course as a wildcat cartridge that was finally awarded a factory chambering in 1987. I know it was chambered in the Remington Classic 700 Limited Edition at one time. Its popularity has waned over the years. It is now immensely popular in Mississippi and should be equally useful all across the country.

Remington originally loaded the .35 Whelen with 200- and 250-grain soft point bullets generating a muzzle velocity between 2,400 and 2,675 fps. The energy rating ran 3,177-3,197 foot pounds. The Whelen kicked at both ends, but was useful for nearly all North American big game.

Most recently Hornady ammunition has brought out a new load in the Whelen. Theirs uses a 200-grain soft point GMX bullet in their Superformance line of ammo. Its muzzle velocity is listed at 2,920 fps, so it is pretty hot. Sighted in at two-inches high at 100 yards, it will hit zero at 200 yards, and nine-inches low at 300 yards.

Some could argue the .35 Whelen is too much gun for white-tailed deer, but such things are usually quite subjective. The Whelen is indeed a thumper, but then my all-time personal favorite for deer is the 300 Winchester Short Magnum. Go figure. Trust me, if the shooters do their part, the .35 Whelen is great for Magnolia deer, as I was soon to discover.

As a gun enthusiast and avid deer hunter, I have rarely relegated my choices to the selections off the rack. The .35 was not available in many "approved" rifle models save the rather lightweight H&R break open version. I had already heard the horror stories from deer hunters that had traded theirs back to dealers for something that kicked less.

So, I not only opted for the Thompson/Center Encore single shot as the basis for my own personalized "primitive" weapon, but I created it from the action up. I ordered the stainless steel frame from one source, and then found a stainless 28-inch, fluted barrel on-line from another source. I did not like the factory synthetic stock that came on the frame, though it is a good one, just not my taste.

I searched and searched until I ran into the Arthur Brown Company which specializes in T/C barrels, stocks, and accessories. They market a beautiful brown laminated wood thumbhole stock and I had to have one.

After the essentials were assembled, I added a Game Reaper one-piece scope mount with integral rings for the Leupold Vari-X II scope, a 3-9x50mm version in a matte black finish. I have to admit that black scope against the stainless rifle and brown laminate stock is quite an eye-appealing package. But would it shoot?

As is common with things at my age, I went to the range to discover the rifle was shooting all over the paper. Hmmm, I forgot to boresight it. So, I took it home, fixed that and started over. Next round at the range I put a whole box of ammo through it for the practice, sighting it in for two inches high at 100 yards. It does kick, but not offensively.

During the opening weekend of the new early primitive season, I took the first deer in camp with the .35 Whelen T/C. It was a convincing hit at a tad over 100 yards with complete terminal effect.

All said then, my trials with the .35 Whelen have been highly successful. I liked the set-up so much, I used it all season long last year. I highly recommend others try the .35 for deer hunting. I think you will find it is a great choice. Check with your local gun outlets and ask to see one. Take your checkbook.

GETTING INTO AMMO RELOADING

With ammunition prices increasing and supplies decreasing, many shooters are opting to get into ammo reloading. On the surface it might sound like a technical project you might not want to tackle or too expensive to establish. In reality, neither is really the case.

Reloading your own ammo can not only become a fun hobby, but a very cost-effective one, too.

I reloaded my own handgun ammo for years, concentrating mainly on simple straight cased stuff like the .38 Special, .357 Magnum, and the .44 Magnum. These were rounds I shot the most, but I also reloaded some hunting rifle ammo as well, including the .243 and 30-06.

If asked what to advise someone wanting to get into ammo reloading for the first time I would highly recommend they purchase a complete reloading set-up kit from one of the major reloading equipment manufacturer/suppliers. The big names here would be Hornady, RCBS, and Lyman. You can easily shop the big catalog companies or likewise on-line to find all you want to know about their equipment, tools, and supplies.

At the same time, for pure neophytes, buy a good reloading manual. These books will walk you through the reloading processes and procedures and also offer reload guides for virtually every rifle and handgun cartridge there is. There will be information on powders, primers, and how to assemble loads properly and safely.

The basic equipment included a reloading press. Next comes the cartridge dies by cartridge designation, shell holders, powder scale, powder measure, primer tool, various hand tools to set up and change out the dies, etc. Other pieces include a lube pad for the brass, funnels, cartridge trays, a hex key set, and other stuff.

Later you will need/want a primer pocket cleaner, brass cleaner, deburring tool, powder trickler, a primer tray and other pieces to make the process easier and quicker. As you research the process, you will learn how to go basic or more deluxe like any hobby.

Naturally you will need to buy brass (or previously fired), primers, powder, and bullets. The proper ones will be determined by the "load" you choose to assemble from the reloading manual suggestions. Follow them to the T and do not deviate from the load details given in the book. The loading sequence will be fully explained in the manual.

THE DEMISE OF THE .40 CALIBER

You'll recall my dad's saying: "That is a solution to a non-existent problem." When the .40 caliber pistol cartridge first saw the light of day around 1989, the excitement seemed well-founded. It's seen a good run, but some say its days are numbered in terms of popularity.

Created by Winchester and Smith & Wesson, the .40 cal. was designed to fit into the scheme of things for the FBI after their disappoint with exploring the 10mm. The FBI apparently wanted the power of the 10mm, but in a shorter package. That idea developed into the .40 S&W.

Using bullet weights from 135 with jacketed hollow points up to 180-grain bullets, the .40 created the power level needed, but also delivered a noticeable increase in pressures as well. This then translated into a pistol round that had considerable muzzle flip and hand control issues for trainees as well as agents in the field.

Law enforcement was pretty quick to adopt the .40, too. After a long trial of trying to adapt its police and troopers to the .40, its excitement appears to be waning. A recent article published by a shooting organization noted that the .40 pistols were harder to learn to shoot well, and required more range orientation and practice. This translated into higher costs for ammunition and the time necessary to get officers proficient in the use of the new .40 pistols.

I have not seen any evidence testing the .40s' effectiveness on the streets or in everyday use by police or federal agencies that took on the .40. Ballistics on paper would suggest the .40 should be a top performer, if and only if the users found confidence in shooting it accurately.

The standard loads produced muzzle velocities from around 1,300 in lighter bullets to 1,000 fps in the 180-grain. Energy rating range from 412 foot pounds up to 524. Using jacketed hollow point bullets was the standard fare, so by discussion the .40 would seem a formidable cartridge.

Still, after years of trials in real use and practicality issues, it seems that the .40s are losing their grip. With the advent of new, better, more effective 9mm ammo, many departments are trading their .40s for 9mms. Some units are even going back to the time honored and proven .45ACP. One report suggested the marketplace may soon be flooded with used .40s.

DILLON'S RELOADING EQUIPMENT CATALOG: A CLASSIC

While I don't normally recommend a product catalog, there are a few exceptions and Dillon's *Blue Press* booklet is certainly one of them. Even if you do not reload your own ammunition, there are enough other shooting products presented in this periodic publication that makes it a vital resource for ammo reloaders, shooters, and gun enthusiasts.

The *Dillon* company is of course one of the premier manufacturers of top flight ammunition reloading equipment. Known simply as "Blue Gear" to many that create their own ammo concoctions, Dillon makes reloading equipment and loaders often referred to as "Progressive" reloaders, because once set up and dialed in, one pull of the press handle essentially performs all of the brass loading steps at once.

Because Dillon Reloaders are automatic-indexing it means the loading technician (you) does away with many of the manual processes required by more conventional ammunition reloading equipment. Also many accessory items are available to add to a basic press set-up to make the whole process seamless. For no other reason than seeing these reloaders, you have to order this catalog.

Dillon reloading equipment just scratches the surface of items listed in their catalog. The index listing is long, including books, videos, ear and eye protection, ammo boxes and bins, Dillon leather, range bags, targets, target stands, firearms maintenance supplies, and over 20 other types of shooting accessories.

BASIC PREPPING ESSENTIALS: WEAPONS

The "other" list includes rifle and pistol magazines, sights, grips, gun parts like springs, tactical lights, specialty leather holsters, Wilson Combat items, cleaning supplies, and much more.

Also included in their catalog book are articles covering reviews of historical weapons, classic firearms, product reviews, and profiles of many different kinds of firearms. The product photography in the catalog is first class, especially including the usual sharing of photos of one of the "Dillon Babes" shown from their annual Dillon Precision Calendar.

What I have found most helpful with a Dillon catalog, even though I do not reload my own ammo, is exposure to shooting products I rarely see in any other venue. This includes old staple products we have known for years as well as new items just coming to the shooting marketplace.

So, add to your list of "to do" items for 2016. Order a Dillon Blue Press catalog. Check out their website at www.bluepress.com to order this publication. You won't be disappointed.

"FLAT SHOOTING" DEFINED

A so-called flat shooting cartridge might well just be in the eyes of the beholder. Still, it is a concept often mentioned in discussions at firing ranges about a particular caliber or cartridge to demonstrate its ballistics, downrange flight pattern and terminal effectiveness. Besides being hot around the campfires, this topic can really heat up some debates as well.

Conceptually, "flat shooting" implies that the bullet from any particular round follows its flight path in as flat as a trajectory as possible. This also implies that the bullet should fly further because the bullet is not wasting flying time going through an arch path on the way to the target.

Flat shooing bullets are also presumed to fly flatter as the velocity increases. Thus, the invent and excited adaptation of the magnum cartridges such as the ever popular .300 Winchester Magnum. A bigger cartridge case holding more powder produces a greater span of velocities and in turn the bullet flies flatter, delivering better downrange efficiencies. But, how much better?

Let's take a look at four popular and well researched big game cartridges. The .270 Winchester, the venerable time-honored 30-06, the .300 Winchester Magnum, and the newer .300 Winchester Short Magnum, my own personal favorite. Do the magnums really deliver that much of a flatter shooting edge over these two popular non-magnums?

Comparing these four cartridges using basic standard bullet types and weights, the comparison here is for the .270 with a 130-grain bullet, the 30-06 using a 150-grain bullet, the .300 WM with a similar 150-grain bullet and the .300 WSM also using a 150-grain bullet. Let's inspect the downrange trajectories in inches at 100, 200, 300 and then 500 yards.

	100	200	300	500
.270	1.2	-0-	-5.7	-33.7
.30-06	1.4	-0-	-6.4	-38.3
.300 WM	1.1	-0-	-5.6	-33.4
.300 WSM	1.1	-0-	-5.4	-32

From these factory catalog ballistics you can evaluate for yourself if the magnums do in fact yield that big of an advantage over the non-magnums. As you can see, in reality there is very little difference even out to 300 yards.

The real practical difference would be of course in the terminal energy of each of these. At 300 yards these in order produce 1,482, 1,782, 2,256, and 2,316 foot pounds. In this regard the magnums excel, which is why they are magnums, as flat shooting or not.

G2 RESEARCH'S RADICAL NEW AMMO

Radical: Favoring drastic or extreme change. That definition definitely reflects the attitude that G2 Research wants to project with its new R.I.P. or Radically Invasive Projectile pistol ammunition. Though relatively new to the ammo marketplace, G2's offerings of several varieties of high impact bullet designs have caught on like wildfire.

"The last round you will ever need." "I've been in the ammunition business for many years and I wanted to create a round that would work well against a home intruder," says Cliff Brown, president of the Windsor, Georgia company. "There are so many stories out there about a woman trying to defend her home and having to shoot someone five or six times and they'd still come after her, we wanted to create an effective one-shot man-stopper," Brown reported to the *Kansas State Collegian.*

The R.I.P. ammo features a unique bullet that is a 100% copper design described as a "petaled-trocar" design. The bullet has the ability to penetrate deep into a target but also fragment in fluid. When the bullet enters the targeted gel or fluid (read animal body/human) the bullet petals break from the base and create individual wound channels as the base continues to push on course.

This penetration factor and petal separation not only delivers a maximum shock wave on/in the target, but an invasive wound pathway as well. For example, their 9mm, 92-grain R.I.P. projectile bullet is designed to create eight copper shards that come from three different angles as they taper to the bullet point. These bullets intend to splinter into eight eviscerating shards and one can only imagine the destructive wound creation upon impact.

R.I.P. pistol ammo is currently available in .380 ACP, 62-grain, 9mm, 92-grain, .357 SIG, 92-grain, .40 S&W, 115-grain, 10mm, 115-grain, and .45 ACP, 162-grain. The ammo is sold in 20-round boxes. Complete ballistics information can be found at www.G2Rammo.com.

Also available from G2R Ammo is their *Civic Duty Extreme Expansion* ammo in 9mm. This ammo offers "exceptional knock down power by delivering an intense energy dump on the intended target while limiting the depth of penetration thus maximizing take down ability with quick and decisive blunt force." The 100-grain load has a velocity of 1,230 fps creating a bullet expansion diameter of one-inch.

Check out G2 Research Ammo on-line today, then get your dealer to order some. I will do a more detailed range report on this ammo in the near future.

HOW TO BUY AMMO CHEAP

One of my hobbies, if you want to call it, that is to work the local monthly gun shows with a friend of mine who is a Glock armorer. While he works on Glocks, I sell accessories, and during the slow times I cruise the 400+ display tables to gauge the gun-selling market, look for new models coming in for sale, and monitor things like ammo supplies and costs. This is all information I hope to incorporate into materials for my readers.

After the last "crisis" of sorts when the Feds were going to stop production of green tip 5.56 ammo, the usual results occurred. First of all, the prices per box and case lots shot back up 20-30+ percent back nearly to where it was after POTUS was elected the second term.

The crowds at the show were lined up a hundred yards deep a half hour before the show opened. Panic-stricken buyers rushed the tables that had ample supplies of ammo. For two hours I watched buyers toting by cases of 9mm, 40 S&W, .45 ACP, .223/5.56 and .308 ammunition.

The best branded stuff naturally went first. Ammo buyers and shooters still know what the good stuff is and still want to buy it virtually at any cost. Dealers quickly sold out of PMC, Federal,

Remington, and Winchester ammo. The crap ammo went next and the real trash sold a little, especially in 7.62x39 for the AKs.

Gun shows are still a good place to buy cheap(er) ammo. Wait until the panic is over. Always peruse big box stores for sales items on the shelf that are not advertised. I just saw 30-06 Remington Hog Hammer ammo for $20 a box. This is not just for hog hunting.

Stop in to the smaller mom and pop gun shops and flash some cash for a discount. Buy ammo in case lots to get a better deal. Even some of the Internet sporting goods sites are now offering free shipping and better prices to generate business. Buy all you can.

Ammo is never going to really be *cheap*. It can be more affordable if you shop around and are prepared to buy when the deals are good, or at least reasonable.

IS BALL PISTOL AMMO VIABLE?

Ball ammo is perhaps the most standard-type pistol ammunition available. So, when the question arises, is it viable, then you simply have to ask "for what?" Ball ammo was designed with a basic full metal jacketed bullet so it would function reliably in a semi-automatic pistol. These round, smooth bullets were created to feed from a pistol magazine flawlessly and enter the chamber without snags or hang-ups. In this regard, ball ammo has performed admirably.

I imagine ball ammo has been the standard military issue since the ubiquitous Colt 1911 came on the scene. It is the round that came with my dad's 1911 in WWII as issued to B-24 pilots. When he brought home his 1911 the ammo still in the original, unfired pistol and steel magazine was ball-type ammo. Only by the time he got home, the bullets had turned green.

Ball, or metal case ammo as it is sometimes called by factory manufacturers, is excellent for general practice shooting whether your pistol is a .380, 9mm, 40 Smith and Wesson, .45ACP or similar semi-auto. Another feature of modern ball ammo is that it is relatively inexpensive as compared to specialty ammo for hunting or close quarters personal defense.

Ball ammo is generally available in rather standard bullet weights, too. The .380s use 95 grains, the 9mms use 115-124-147-grain bullets, and the .45ACP standard bullet weight is 230 grains. The .40s use a 165- or 180-grain metal caseloads.

Now, viable you say? Is the ball ammo the best choice for hunting or self-defense, or law enforcement? No, it is not, especially with the advent of so many new bullet designs for these purposes. Is it the load that many shooters keep stocked in their magazines a lot of the time? Yes, for expense and reliability issues, ball ammo is still tops.

So, if you shoot a lot and want to be proficient with your pistol, then ball ammo is the way to go. If you get accosted in the parking lot, or have some other unfriendly confrontation, will ball ammo get the job done? You bet it will. I'd certainly call that viable.

IS THE .38 SPECIAL STILL VIABLE?

The .38 Special is not the lightweight most people think it is. And small revolvers typically chambered in this round are not intended just for small-handed people or women who never learned to shoot properly. Maybe it is time to rethink the value of this formidable round.

Originally tagged as the .38 Colt Special, this cartridge was developed in 1902 by Smith and Wesson for the Military and Police Model revolver. It was created to replace the older .38 Long Colt used unsatisfactorily by the Army. Colt brought out their new handgun version for the new .38 Special in 1909. Please see all the old black and white gangster and police movies.

Quoting from the landmark gun text *Handgun Cartridges of the World* by Frank Barnes, "The .38 Special is considered one of the best balanced, all-around handgun cartridges ever designed. It is also one of the most accurate and is very widely used for match

BASIC PREPPING ESSENTIALS: WEAPONS

shooting." Plus, any .357 Magnum revolver will also shoot the .38 Special without prejudice.

The typical and perhaps original load was the ubiquitous 158-grain, soft lead, either round nose or flat wadcutter bullet. Today, .38 Special ammo comes in dozens of bullet weights and designs from 95 grains up to 200 grains. Bullet types include jacketed hollow points, wadcutters, lead round nose and other modern specialty self-defense loads.

These include loads such as the new Hornady 110-grain, FTX (Flex Tip) bullet at a muzzle velocity of 1,010 fps with a muzzle energy of 249 foot pounds of energy. This is just one example. Hornady also has a .38 Special load using their 125-grain XTP (Extreme Terminal Performance) bullet that cranks out a MV of 900 fps and a ME of 225 ft. lbs. Virtually every ammunition company loads the .38 Special so it is widely available worldwide.

The .38 Special is most commonly chambered in the five- or six-shot revolver type handguns. These come in many fashions, from snub nose pocket models, to full-sized revolvers with four- to six-inch barrels and even some older models with eight-inch tubes. Most of these handguns come with "open" sights or really just a flat top with a sighting groove on the top of the frame and a half-moon front sight or other aiming fixture. There are adjustable sight target models as well.

The .38 Special is still viable as a self-defense, prepper, or close quarters survivalist weapon for pocket or waistband carry. The .38 Special is just as special today as it was in 1902 when it first appeared.

CHAPTER 3
AR RIFLE ACCESSORIES AND STRATEGIES

AR MAGAZINE STRATEGIES

If you own and use an AR rifle platform then you need an AR magazine strategy. Have you ever seen so many options for ways to feed a rifle than with the vast varieties of AR magazine configurations? You might think you were being even more confused than just trying to pick all of the AR rail accessories on the market these days.

It does not matter the reason you use your AR either. Maybe you just like to punch paper targets or bleeding zombie forms. More and more hunters are using ARs for varmint hunting, predator control, or even larger game hunting like whitetails or even bigger game. There are multiple law enforcement applications, professional security applications, or even home or ranch security and protection. AR use has multiple purposes and thus the fire control systems need to be managed, too.

Obviously an AR rifle magazine is the ammo-holding device/fixture that is used to feed ammunition into the firing action of the rifle. These cartridge-holding implements come in a variety of configurations, capacities, construction materials, colors, and other design features and functional differences.

Typical factory "mags" are either of a metal or polymer manufacture. There are AR magazines made of steel and even more commonly aluminum. You'll notice the ever so slight differences in weight when you handle and load them. Steel mags can rust and thus require more attention to maintenance issues. Aluminum magazines can bend or dent more easily, especially around the crucial feeding lips at the top of the magazine opening where ammo rounds are inserted.

Under normal everyday use by the average AR shooter a magazine could last indefinitely if maintained and cared for, properly stored and carried. LE users and others like military applications are more likely to see cause and affect damage to magazines beyond reliable use. Care for your magazines, but inspect them regularly and toss ones that fail to function properly.

Today there are many commercial AR magazines made of exceptionally strong and nearly maintenance-free polymer materials or hard-like ABS plastics. The most prolific brand on the market today that readily comes to mind are the magazines made by *Magpul*. These are now made in the great state of Wyoming, having moved from Colorado for political reasons.

Magpul magazines are well-made, feature a snap-off dust cover and are virtually indestructible. They can also be fitted with factory "snatch" finger grabbers or easy magazine extraction from the rifle. These come in 20- or 30-round capacities and are available in various colors besides standard black.

AR magazines come in different holding capacities including 5-9-10-20-30 and 40 rounds in the usual single stack self-contained types. There are also 60- and 100-round magazines on the market like the *Surefire*, which are of a double-stack type. Rotary feed magazines like the *Beta* 100-round dual cylinder magazines are popular, too if heavy firepower is needed. Another rotary choice is the *X-Products X-15* magazine. These are high-quality and reliable.

AR magazine storage is not rocket science, but some common sense needs to be used to keep them in dust-free environments secure from prying eyes and fingers. They are easily kept in storage "crates" like *Plano* snap lock gear boxes or *Case-Gard AR Magazine* tote-ammo-type boxes. I like the Case-Gard boxes with the foam magazine inserts to keep things organized, safe, secure, and clean.

GET A SLING WITH CLING

Sometimes a topic seems so simple and such a no-brainer that we laugh at the title. This could be one of those ideas. I mean a rifle sling is a rifle sling, right? All they were designed to do is tote a rifle over the shoulder. What could be so complicated about that?

Ask yourself this one simple question. Are you happy with the rifle sling you have on your primary hunting rifle right now? Do you even pay much attention to how well it functions at the job it was created to perform? Does it bear the weight of your hunting rig with scope and being fully loaded with a reasonable level of comfort and security? Has it ever slipped off your shoulder causing you to drop your rifle? Yeah, I thought so.

Certainly, a rifle sling should carry the weight of your hunting rifle securely on your shoulder. Most hunters simply throw the sling-installed rifle over the shoulder and hike on, or up or down. I suspect most hunters pay little attention to the performance of their rifle sling until it starts digging into their shoulder and collar bone after they have shifted it from one shoulder to the next back and forth several times. Are there better approaches, or better materials to make rifle carry more comfortable and stable?

Personally, I hate a rifle sling that constantly slips off the shoulder, causing me to grab the rifle quickly or otherwise find it lying in the mud. A few years ago I was coming down the ladder of a 16-foot tripod stand with my unloaded rifle slung over my shoulder. During the process of changing hand grips on the ladder rungs and body movement shifts coming down the stand ladder the rifle sling came off the shoulder and escaped my grip in a nanosecond.

BASIC PREPPING ESSENTIALS: WEAPONS

My favorite Browning deer rifle dinged down the metal ladder and landed barrel first into the muddy ground. The rifle stuck perfectly vertical down with the barrel into the mud about eight inches. With great despair I inspected the gun only to find a small scratch on the scope, but no dents. Back at camp, I scrubbed the barrel thoroughly with a brass brush and re-lubricated it, then I wiped down the whole rifle rig.

Then I unattached that lousy nylon sling and threw it away. At that moment I initiated the search for a better sling that had some gripping power to secure my rifle against my hunting clothes and body. And just in case you are curious, I too was surprised the rifle continued to shoot ideal groups at the target range without any changes after the drop.

A standard nylon sling is a poor excuse for an accessory to carry a heavy hunting rifle in my opinion. Now on a lightweight .22 rifle like a Ruger 10-22 or something like that, a simple nylon sling works. I cannot argue against a nylon sling being tough and durable, but it is also slick and easily slips off the shoulder. It is also somewhat abrasive and uncomfortable, especially for those common slings that are only an inch wide.

Narrow leather slings also fit into this same category to my way of thinking. Now I do own and use a couple of two- to three-inch-wide leather slings, one with a wool-fleece-type facing that carries a rifle well and is pretty good about staying in place on the shoulder, but there is no stretch to it. Some hunters and target shooters like the old-fashioned Army-type leather sling with multiple holes and a claw-type hook adjustment. It's just not for me.

The newer "rubber-type" slings like a *Quake Industries Claw* sling seem to serve pretty well. I have had some issues with them once they get wet in a light rain or mist. They are easy to grip in the gun-shouldered hand though for extra control, which is how I carry a rifle anyway. The polymer pad is non-slip and stretches for added comfort during long carries.

Some fabric slings made of cotton, nylon synthetics, or polyesters that feature a gripping surface sewn down the center of the sling work pretty well, too. I have one of those I use solely on my Remington 11-87 turkey shotgun and it does well.

Of all the slings that I have tried over the years and now use almost exclusively on all my regular bolt action and AR hunting rifles are those made of elastic neoprene. The brand I buy is *Vero Vellini* in different widths and colors. The end tabs of these slings are over-sewn with leather for extra durability and appeal. Bar none, these neoprene slings are the most comfortable I have used in all kinds of hunting east and west in both good and bad weather. They stretch and give with any body movements and can carry a 10- to 12-pound rifle rig comfortably all day.

Sling attachments come in all types, too, so let me give you my thoughts on that. Unfortunately, my all-time favorite *Uncle Mike's* one-inch sling attachments without the screw-down lock are no longer made, as far as I can tell. I like the old-type push-button sling swivel that rotated the end piece to the side so the bar could be slipped out of the swivel stud on the rifle.

I always installed these swivels in opposite directions so each end of the sling "pulled" against each other. I never had one come open or off inadvertently. If I could find some more of these old-style swivels, I would buy them in a minute.

As it is, the newer-type swivels have the locking screw-down security caps. There really is nothing wrong with these, though occasionally I notice they do unscrew themselves with use. These screw locks can be fidgety to unlock with wet hands or gloves.

Also, I prefer metal sling swivels to anything else. They make ABS plastic and hardened nylon types, but I have had them break and warp so I no longer use them. Metal ones do require inspection for rust. A simple brushing with a bronze/brass gun cleaning brush and some oil fixes that in a couple minutes. Also adding a couple drops of oil periodically to the swivel installed into the rifle stock stud will prevent them from squeaking.

So, you thought a rifle sling was a simple deal, huh? Well, they are if you find one that works for you and provides all day carry comfort and rifle security. Trust me, you don't want to have a hunting rifle rig slip off your shoulder on an expensive western elk hunt and bang the whole thing off a big rock. Been there, done that.

THE *SUREFIRE* HI-CAP AR MAGAZINE

Whatever SHTF scenario plan you have devised, whether it is a classic well-prepped Bug Out escape or an Alamo standoff on the home front, more than likely there is a firearms component written into the flowchart. Having reviewed many forums for some time without being compelled to engage any longer, I do know that firearms and their use are a huge concern for a lot of preppers. The fears and paranoia are justified.

Many preppers have settled on a weapons offensive and defensive system that they hope will prove effective if ever pressed to deploy it. I suspect the majority have selected at least one component of a weapons approach to be fulfilled by one of the many options of an AR rifle. Given everything we know or can learn, the AR proves to be a pretty darn good choice for a SHTF rifle, all else being equal. There are other viable choices, too, but there are more AR options, accessories, and ammo choices.

When it comes to feeding the AR via lower unit magazine insertions, the traditional modes are mostly of the 20- or 30-round standard steel, aluminum, or polymer magazine types. Without a doubt these work, some better than others. So be careful what you buy and test out thoroughly anything you plan to rely on to defend yourself and your family. Buy as many as you can afford to stock and keep them loaded.

For a company with lots of bright ideas and a well-known reputation for their line of high-quality flashlights, it is even more refreshing to see such an outfit bring out something like their new

AR high capacity magazines for the common 5.56/.223 round. The use of these 60- or 100-round magazines are growing in use in military and law enforcement applications, so why not for survival applications as well? They can solve a lot of problems for preppers.

Known simply as their 60-/100-round High-Capacity Magazines Models MAG5-60 and MAG5-100, these Surefire magazines certainly have a different look and feel coming out of the plastic slip package from the factory.

I have had one of the 60-round mags for field tests and it has a quality feel as one handles it. Upon examination, the hard-anodized aluminum body is a solid, well-built piece of equipment. The seams are smooth and well-finished-out. Surefire uses 6000 series aluminum in their magazines.

The spot welds on the front and rear body parts are clean and smooth. The floor plate is well-finished as well. These Surefire magazines are totally compatible with all M4/M16/AR15 rifles and variants.

These magazines have earned a reliable and durable reputation in fighting fields all around the world. They feed smoothly and reliably due to their optimal geometry, non-binding coil springs, and nesting polymer follower. Lubrication is not required in the use of these magazines. Being reliable and low maintenance are two good reasons for having Surefire mags in the Bug Out gun bag.

Surefire magazines can be ordered on-line at their website at www.surefire.com. The 60-round mag retails for $129.00, while the 100-round version goes for $179.00. Shop the net to look for better pricing from some sources.

This may be a personal preference issue for any AR user. The typical AR configuration with either a traditional fixed buttstock or a six-position collapsible stock handles about the same with a standard 20- or 30-round magazine installed for carrying, shooting, security, or hunting purposes. The weight and ease of firearm manipulations are roughly the same. Most AR users are familiar with this especially out in the field

Now add on a single magazine that can hold 60 rounds at a time. Believe it or not, the weight pulling down on the rifle actually enhances rifle control. Sure the 5.56/.223 doesn't muzzle jump, but some models of the AR can be fairly light. Having this little extra bit of heft can make the rifle more stable for more accurate shooting. So equipped, they work great off shooting sticks as well.

I often see AR hunters with two regular mags clamped together with various kinds of magazine coupler devices. In my opinion this is awkward and far less suitable than a single high capacity mag like the Surefire. The major point of course is that the one mag holds up to 60 rounds at a time without having to change out a magazine. That's a plus in my book.

So, if you feel inclined to load your AR prep rifle at daylight in the hopes that the one mag will get the job done for an extended period of time, then this Surefire High-Capacity Magazine is the way to go. If not, then you could always get two of them or the 100-round model. Either way, the Surefire magazine will keep you reliably in service for quite a while. Think this over as you plan further for equipment buys for your prep, home, ranch, business, or personal protection.

AR BUTTSTOCKS - FIXED OR SIX-POSITION

AR rifles or MSRs (Modern Sporting Rifles) are well-known for the flexibility of their platforms. These rifles are the most adaptable to the widest range of factory and aftermarket accessories of any gun design ever. An AR owner can virtually customize their own rifle to add dozens of features to suit their own shooting needs, requirements, and wants. The list is endless.

One of the features an AR buyer can pick right off is the style of buttstock they prefer to use. Factory rifles come with one of two types of buttstocks as a general rule. This will be either the fixed stock or one that is adjustable for several overall length positions.

Usually these adjustable stocks are known as six-position or collapsible stocks. Technically these are known as the A-2 fixed stock or the M4 adjustable stocks.

The original Stoner ARs and later the mass-produced Colt ARs of the Vietnam War fame all arrived from the factory with A-2 stocks. Later a more or less combat/special forces version was developed to be called the M4 models. They both had their place on the battlefield.

Today, the fixed stock versions coming out of the many AR manufacturing factories are models best suited as hunting or target shooting rifles. The fixed stock is sturdy and provides a consistent overall balance to the rifle. There is no give or take and no characteristic rattle that is often found with the M4 type.

The A-2 stock also provides a consistent cheek weld when shouldering the rifle for hunting or punching paper targets. Law enforcement professionals that have trained to use ARs as sniping rifles also tend to prefer the fixed stock for better rifle control on downrange targets.

The six-position stocks are extremely popular for a wide variety of reasons, most being the flexibility of changing the length profile of the AR rifle for tighter shooting conditions such as close quarters combat or defensive shooting. The M4 stock when collapsed down also permits storage and carry as a compact package as well as for concealment.

M4 stocks are adjustable by a variety of means using levers and releases of different kinds, depending on the manufacturer. These buttstocks can also be found with other features including various sling attachment modes, battery storage and such.

When picking your AR try out both types at the dealer to see which buttstock style fits your needs best.

BASIC PREPPING ESSENTIALS: WEAPONS

PRACTICAL CARRY CASES FOR AR RIFLES

A well-decked-out expensive AR rifle setup demands a good carry case. ARs are not fragile, but they need protection from everyday carry outside the house. Environmental elements including precipitation like rain, sleet, or snow as well as windblown dust, dirt, mud, and just plain ole grime will attach itself to an exposed rifle. It just makes sense to have a good case.

Yeah, I know your rifle is going to collect this junk when slung over the shoulder anyway, but until it is deployed in use, it needs to be inside a good protective case. These not only keep the natural elements at bay for a while, especially sitting out in camp, on the tailgate of a truck, or on the rack of an ATV/UTV, but they reduce incidental knocks, bumps, and scratches that can potentially impede the function of the rifle, optics, or other accessories installed on the rifle.

What constitutes a quality AR rifle case? First ask yourself the primary purpose for the AR. If it is a SHTF or prepper rifle, then you might want to maintain a lower profile with a gun in the vehicle or whatever. If it is a hunting rifle, then not so much concern about keeping it discreet, but still protected and unnoticed.

If you want the rifle to be non-discreet, then shop for a plain black zippered case with good padding. Nylon material will wear the best and shed rain droplets, spills, and can easily be brushed off or cleaned with a damn rag. Go with the hand grab handle and skip the shoulder carry strap and outside Velcro® pockets.

Converse to this type of case, if you hunt or perhaps are law enforcement, or security, then the cases with heavy padding inside and additional gear pockets outside are the way to go. Buy a case with a good, sturdy zipper that goes both ways. Well-designed cases made just for ARs and like rifles will have built-in tie-down straps inside to keep the weapon from shifting in the case.

Color-wise, I think black is the best bet. There are AR cases now that look like sports bags or take-down cases that appear to be tennis racket bags. I can see that working. Whatever gun case you get, make sure it protects the rifle.

VARIANT CHAMBERINGS FOR THE AR PLATFORM

Believe it or not, the AR-15 rifle platform comes in a wide variety of cartridge chamberings besides the .223/5.56. I highly suspect that the standard .223/5.56 chamber comes in probably 90-percent-plus of the ARs manufactured. However, this should do nothing to deter lovers of the AR configuration to actively try out other viable chamberings in an AR rifle.

To date, and perhaps this statistic is outdated, but the AR-15 rifle currently can be acquired or fitted with four different rimfire calibers including the .17HMR, .17 Winchester Super Magnum, the .22 Long Rifle, and the .22 Winchester Magnum Rimfire.

There are 17 centerfire calibers available in the "inch" measurement category in an AR-15. The most well-known would be the .17 Remington, .204 Ruger, .243 and .25 Winchester Super Short Magnums, .30 Carbine, .30 Remington AR, .300 AAC Blackout, .300 Whisper, .450 Bushmaster, .458 SOCOM and the .50 Beowulf. If you are not familiar with some of these chamberings then run a *Wikipedia* search on them.

Next there are the centerfire calibers in the metric measurement category. There are an additional 10 chamberings in this group. Popular choices here include the 5.45x39mm (aka the AK-47), the 6.5 Grendel, 6.8mm Remington SPC, the 7.62x40mm Wilson Tactical, and the 9mm Parabellum and 10mm auto pistol rounds.

So total we are talking an additional 31 chamber options in the AR-15 Platform. In many cases all that is needed to change from the .223/5.56 to another caliber is to swap out the upper units on a standard lower unit. Sometimes in cases like the .300 AAC Blackout and .300 Whisper, all that is needed is the upper with the use of the exact same .223 magazines.

In other cases a complete new rifle is needed for the round to function as a complete rifle. These might include the 6.8 SPC and the .458 SOCOM. I certainly like the idea of just buying an upper unit and quickly slapping it on top of an existing lower and going to work.

Of all these additional chamberings, I have shot many of them, but personally use the 6.8 Remington SPC in a Ruger AR and the .300 AAC Blackout (along with the .300 Whisper which are basically interchangeable). I use the 6.8 on wild hogs, and have successfully used the .300 Blackout on white-tailed deer.

AR MAGS - METAL OR PLASTIC?

This is a subject of contention that never occurred to me until recently. Another prepper friend of mine brought up the subject and we had quite a lively debate about it. Everybody has their own preferences about such things like what they want on their hamburgers. I am more like the guy on the TV ad that just wants things to work. I am all about things working right, especially firearms magazines.

Without conducting a comprehensive review of the current marketplace, it seems most AR magazines are made of steel, aluminum, or various kinds of thermoplastics. We all know steel can rust, but otherwise they are tough, but heavier in general. Aluminum mags carry lighter, but can dent easily. Serious dents can yield them unusable. Many AR users preferred metal mags years ago, but I think the tide has turned on that.

When plastic AR mags first came on the market, some of them were pretty rough in terms of fit and finish. As time went on the whole field of mold manufacturing has improved dramatically and thermoplastic mags today such as those by Magpul® are highly respected on the battlefield as well in the hunting fields and woods around the world. They just work.

Over the past 20+ years I have used a wide variety of AR magazines and I have to say that my luck with all types has been very good. The first ARs I worked with all had aluminum 5.56/223 magazines supplied from the factory and not one ever failed. Of course, I was not in combat either or using these rifles in dust storms, swamps, or other particularly tough environmental conditions. Under normal hunting use, care and maintenance they have all worked well.

If anything tends to break down in a weapon's magazine it seems to be due to either the follower part or the mag springs weakening over time and failing to provide enough pushup force to make the magazines feed correctly.

I have on only rare occasion taken the floor plate off an AR magazine. The rare exception has been when I added Ranger Plates to my Magpul mags. When I do that, I wipe the springs down with a lightly oiled cloth since it was convenient.

So, buy mags that work of any material. Today, I think they all tend to work pretty well.

AR RIFLE PLATFORM RAIL OPTIONS

The single biggest attractive feature to the AR rifle platform is its ability to be customized. Big in this regard is choosing one of the many rail optional accessories available in the marketplace. Many ARs come from the factory equipped with a variety of rail formats already installed. Other AR owners can shop a wide variety of aftermarket accessory suppliers to find just the right combination of parts to outfit their rifle to suit their own needs.

There are now two basic forms of rails for attaching accessories. The most common and widely available model comes with Picatinny ridges and slots. The name Picatinny comes from the *Picatinny Arsenal* in New Jersey. This military arsenal was charged with evaluating rail systems for adaptation into military use via the Mil-STD-1913.

BASIC PREPPING ESSENTIALS: WEAPONS

The military specifications for rails for the M4 AR rifle were adopted in 1995.

The Picatinny rail accepts Weaver-type attachment designs to fit a threaded rod or square pin with end threads down into the rail slot, then a rail grabber end part is screwed down tight with a bolt nut or knurled knob to hold the Weaver attachment mount in place. These Weaver-type attachments are designed into the various accessory items to be mounted on the rail.

The other type of rail attachment format is known as the Key Mod or Key Mod Handguards. This design offers a series of mounting holes with lock-in slots. The mounting accessories will have matching studs to fit into the Key Mod holes, and are then slid back to lock in place. This system is becoming more popular, but is not nearly as common as the Picatinny rail type.

AR rails can be two- or four-sided, meaning there can be mounted handguards covering the barrel on four sides, right, left, top, and bottom, or other configurations. The four-sided rails are the most common. Handguard sides not being used to mount accessories can be covered with rail covers to protect the hands from the sharp rail edges.

A myriad of accessories can then be easily installed on the rails to suit the user. These can include scope mounts, red dot sights, BUIS sights, flashlights, lasers, sling attachments, bipods, vertical grips, hand stops, and other accessories. The best strategy is to customize your AR platform with useful items, being mindful of the weight you add.

AR SCOPE MOUNT SYSTEMS

Crucial to an accurate pinpoint shooting AR is the scope mount. You have spent major bucks on a good AR rifle setup and then you stretched the budget for a quality optical scope to set the rig up just right. Now is not the time to go cheap on the scope mount.

AR rifle platforms by their very nature are the most accessorizable (is that a word?) weapons systems ever designed. If yours has a Picatinny rail on the top, then you have virtually unlimited options for adding sight systems to the rifle. The options are also many for putting a high-quality, secure scope mount on the rail.

Options for scope mounting on an AR are basically of two types. These are single mount fixtures or double mounts. Obviously the single mount is a one-piece mount of either aluminum or steel. Take your choice here picking weight issues over the inherent strength of steel.

I tend to lean slightly toward steel mounts as I have had less problems with them. Steel can and does rust of course, if you do not maintain it. Aluminum does not rust, but it does scratch and mar easily. I have also had some issues with screws turning out in aluminum-tapped threads. I prefer steel mounts with the blue screw lock glue on the threads.

One-piece mounts tend to be tricky in terms of scope heights and setting forward eye relief. The best part about a Picatinny system is that scope mounts can be moved back and forth in the rail grooves to adjust. You may find with some scopes that an extended mount works best. The single one-piece AR mounts by Nikon are already set up this way. They must know the issues. Other high-quality one-piece mounts are made by GG&G in Arizona. These come in bolt-on or with adjustable quick-release levers.

There are many brands of two-piece scope mounts that fit the Picatinny rails. Some of the ones I have used with good success are GG&G, Leupold, and Midwest Industries. These also come in cross-bolt configurations with heavy nuts to tighten with a socket and socket wrench, but use care in the torque you apply. I have never twisted one off, but it could be done.

AR SLING ATTACHMENT OPTIONS

AR rifles can be carried by a sling in a wide variety of options. There are one-, two-, and three-point sling attachment variations depending on what the rifle user wants and is most comfortable with for their modes of use.

As with the huge variety of factory and aftermarket accessories, there are also multiple options for attaching a sling to an AR rifle. Many ARs come from the factory with more than one way to strap a sling to the rifle, but aftermarket accessories have really broadened the choices.

The typical sling attachment setup on a basic AR is a front sling loop attached via a screw/bolt under the front forearm. The rear sling loop is a metal part that projects down from the buttstock. A common cotton or nylon one-inch sling is simply looped through these two attachment points, looped through a metal sling adjustment slide and adjusted to length properly for shoulder or cross-chest carry. This is essentially a two-point attachment.

With the use of various attachment accessory parts via GG&G, Midwest Industries, Blackhawk, Troy, Daniel Defense and other's parts installed behind the rifle action on the receiver extension tube, slings can be looped in, snap-locked on a ring, or connected via push-button QD swivels. This option becomes single-point attachment.

These allows the sling to be hung over the shoulder, across the chest, or other modes that allow the rifle to hang free, suspended down basically under the arm for quick deployment. A rifle carried this way also permits the rifle to be quickly "dropped" for a quick draw of a sidearm.

The push-button attachment points sometimes come stock on factory rifles and are very popular. I have converted nearly all of my ARs and like rifles to the push-button releases QD swivels. They are quick to snap in or release and are very secure so long as you make sure the buttons are firmly locked into the female insert. These can also be configured for two-point attachments as well.

Three-point attachments are more complicated and seem designed for law enforcement or military in a patrol mode where the rifle is held strapped across the chest. I have tried these once and found them more difficult to rig up and especially to adjust quickly.

So, when you get ready to sling up your AR be sure to check out all the options.

CROSSHAIRS OR RED DOTS FOR MSRS?

A healthy debate continues over the selection of an optical option for AR-type rifles. Really, this does not have to be a debate, as one type of scope is not really *better* than the other, it just depends on the purpose and use of the weapon. And I just throw out the "AR" rifle as a sort of generic reference to defensive type rifles including ARs, AKs, FN-FALs, M1As and any such rifle in this category.

Admittedly I was somewhat biased against red dot luminary type sights as a hunter until I was hosted by Smith and Wesson and EOTech on an Oklahoma deer hunt a couple years ago. I really learned firsthand the value of an electronic-type optical sight with the EOTechs. I guess I was a bit too old school with the use of traditional optical riflescopes.

Of course, in the final analysis there are proper places for the use of both types of scopes on all kinds of firearms including AR-type rifles for hunting or defensive work. With the popularity of AR rifles these days (or MSRs, Modern Sporting Rifles, as we have been taught to be "PC" in calling them such), for hunting varmints and deer-sized game, you will see plenty of these rifles with conventional scopes with one-inch or 30mm tubes, both fixed and variable power scopes. I use both on mine.

Typically we might think of the red dot (or green or amber in the case of a Trijicon) sight as a "combat"-type sight relegated to short range work, but this is no longer the case. Many of these dot sights can be expected to provide ample optical accuracy out to hunting

BASIC PREPPING ESSENTIALS: WEAPONS

ranges of 200 yards or more. For survivalists and preppers they still apply to close quarters defensive work, too.

MSRs are very adaptable to traditional optical scopes. With the vast array of scope-mounting options for these rifles, virtually any scope even with huge objective lenses can be mounted. Use care in their mounting position, noting access to charging handles and ejection issues. The nice part with the Picatinny rails is that mounted scopes can be easily moved back and forth to adjust to the user's preferences.

Red dot opticals provide some new exciting options for MSR users, while the conventional scopes continue to offer quality target acquisition.

CHAPTER 4
SHOOTING GEAR AND TACTICS

CALDWELL'S DEADSHOT CHAIRPOD

Every so often a product comes along that makes one wonder what took them so long. Caldwell Shooting Supplies, part of the Battenfeld Technologies, Inc. group out of Columbia, Missouri makes some great products for shooters, gunsmiths, gun enthusiasts, and hunters. You have to shop their website sometime to see all the products they have available at www.BTIbrands.com. You will quickly recognize some of the top brand names in the shooting accessory industry.

Their latest conceptualization is called the DeadShot ChairPod. This setup is a seat and shooting support all combined in one product. One has to experience it through the process of assembling it out of the box to fully appreciate its features and design.

As with all Caldwell products I have used in the past, they come well-packaged in one shipping/sales floor display box. How they do that alone is a marvel. Once you start pulling out all the pieces and parts, and the main components to lay them out on the floor, one begins to get a sense of how it all works together.

Basically this is a fold-out chair like ones similarly used in ground blinds. The back folds down over the four legs and there is a backpack strap system installed for toting the whole rig to the field for use in a ground blind or on the shooting range. It could easily be used behind an improvised homemade ground blind as well.

The four legs each have a spread-out foot that resists sinking into soft dirt or mud. However, if you use this ChairPod in a really wet environment inside a ground blind with no hard floor, I highly recommend putting down a piece of plywood to help support the seat. This could help keep the shooter's or hunter's feet drier and warmer up out of the muck as well.

Attached to the front of the chair is a bracket that holds the Swing-Arm Post that supports the entire Shooting Rail Assembly with the front and rear gun-holding support forks. There are way too many adjustment features on the ChairPod to explain, but understand when sitting in the seat with your rifle or crossbow in the support forks that you can fully adjust the swing arm height, the setup shooting rail length to handle any kind of rifle, fork height, and everything locks down tight via screw knob adjustments or extension locks like those on camera tripods.

On top of those multiple adjustment points, the seat is fully rotational 360 degrees. The seat is a comfortable fabric mesh which allows for air flow and some soft flexibility. I believe it would be easy to sit in this ChairPod seat for an extended period of time without fatigue setting in. The seat width too is ample for us larger guys with bigger hips.

When the hunt is over, the whole rig can be disassembled in part and folded up for carry using the backpack straps attached on the back of the seat. The Swing-Arm Post slides out of the bracket under the chair and the "tube" fits into two rubber bar holders attached to the side of the seat frame. Then the shooting rail side sits down into two rubberized cradles fitted into the end of the chair seat frame bars. The whole process is more difficult and time-consuming to describe in writing than it is to do in real practice.

Other features of the ChairPod include the spacing between the twin Shooting Rails which is open. This allows for the easy use of an AR-type rifle permitting an extended magazine to fit down in between these two rails. Though the ChairPod looks to be most compatible with a long gun rifle or even a shotgun, the unit will also easily adapt to using and shooting a crossbow. The soft rubberized front and rear holding forks hold any weapon securely.

Naturally the whole functional purpose of the Caldwell ChairPod is to provide benchrest accuracy from a portable shooting platform that can be used anywhere at any time. The 360-degree swivel chair feature and the fully adjustable gun-rest system allows for the maximum range of movements for the shooter and hunter. The whole of the adjustments permits the precise alignment across the entire range of possible shooting angles, up, down, or swinging right or left of the seated user.

The Shooting Rail setup keeps the rifle or crossbow in the ready position at all times. This lets the shooter-hunter also be hands-free to use binoculars or a rangefinder or to take a drink or use a camera without disturbing the weapon in the fork cradles. The rifle or crossbow can be adjusted perfectly to the shooting position desired by the ChairPod user. The shooting rail slides in or out to handle even a 26-inch rifle barrel or a carbine-length long gun.

The frame, seat, and leg components are all metal with a durable powder coating in a neutral tan color. The Shooting Rail and fork parts are aluminum for lightweight construction. Most of the adjustment parts are tough ABS-type synthetics. The entire package weighs 22 pounds and will support a user up to 300 pounds. In layman's terms this Caldwell DeadShot ChairPod is not cheap junk.

With deer hunting season coming up in a couple weeks, I fully intend to put this ChairPod into full practical use in a pop-up ground blind I already have ready to use near a planted wildlife food plot.

Setting the seat back into the blind will give me maximum flexibility of a full range of left to right shooting positions with the height perfectly adjusted to my seat comfort and shooting angle.

What I look forward to is the stability this chair unit is going to provide in helping me to make precise shots at any possible range a deer might offer me.

If this Caldwell DeadShot ChairPod sounds like something you can use, then I highly recommend you acquire one from your local dealer as soon as possible. I think you will be highly pleased with the quality and the adjustable flexibility this portable benchrest platform will provide.

LETHAL PRECISION

This treatise title I am borrowing from a marketing launch by Winchester Ammo for their line of excellent *Ballistic SilverTip* ammunition. I feel that is sort of "OK" because I use their ammunition exclusively in my No.1 deer rifle, a Browning A-Bolt in 300 WSM I bought back in 2001. The minute a local dealer got one of the very first 300 Winchester Short Magnums in stock I was at his counter. Turns out it has been one great rifle.

Talk about "lethal precision!" That rifle, scope, and ammo combination can be attributed to taking 14 white-tailed deer, 12 of which were bucks in 15 shots. I tell everybody that the sun was in my eyes on that one missed shot, also at a buck.

By hunting rifle standards this older Browning can really print some pretty good groups at the deer camp 100-yard rifle range. I mounted a fine Leupold VX-3, 3.5x10x50, matte finish scope with the lighted reticle. This is one terrific hunting optic and with the 50mm objective lens you can squeeze another 15 minutes of light out of any day, even the dark, overcast ones.

I don't know if you believe in metallurgy chemistry or not, but this rifle-scope-ammo combo has made a believer out of me. I have hunted with dozens of bolt action rifles over the past fifty years, but this one is the best hands down. That includes competition in the ranks from a Remington Limited Edition Classic 700 in 300

Weatherby. That rifle has accounted for both of my elk, one at 400 yards and another at 25. With another Leupold scope it has been a real tack driver, too.

All that said to start the discussion about what exactly is an acceptable accuracy from a rather common, factory, off-the-shelf bolt action hunting rifle? This question is among the top 10 I get every season during hunter education classes, seminars, Facebook, or email inquiries. I think maybe it is something too many hunters worry too much about. But let's examine the issue.

Being on the edge of old-fashioned, I still use the accuracy benchmark of sighting my hunting rifles in at three-inches high at 100 yards. Group-wise, if I can consistently print three shots in three inches or less, then my range work is done. I reconfirm that "zero" before every hunting season.

My second personal acceptance standard is putting 5-10 shots inside a 10-inch paper plate at 100 yards. Pretty liberal huh? In fact, many hunters I talk to just laugh at that acceptable level of accuracy. But why, a whitetail's kill zone just behind the front shoulder for a killing heart and lung shot is really no bigger than that paper plate. I have never failed to kill a deer if my bullet arrives within that "boiler room" zone.

Beyond 100 yards? I'm sorry, folks, but I calculate the arrival points of my bullets with my loads using either a computer software ballistics package, or I look up the general standards of where exactly factory loads are intended to hit at various ranges out to 300 yards, knowing full well I am not ever likely to shoot at a deer at those ranges. I might, but I would think on it hard and as long as I have to, to make such a shot. I am not a long-range game shooting advocate. With my old eyesight I can hardly see a deer standing beyond 300 yards, much less shoot one.

So, when you buy a new hunting rifle, regardless of what kind of firearm it is or the cartridge round stamped on the barrel, do what is right to make it hit the target. This goes for those old deer rifles

BASIC PREPPING ESSENTIALS: WEAPONS

already in the gun rack, too, especially if you change out a scope or change to a new brand, type, or bullet weight.

First, read the owner's manual. Thoroughly clean the bore from the chamber end, not the muzzle. Brush down the bolt, extractor, and all else with a toothbrush to remove any debris from the manufacturing process and packaging. I often find little bits of package Styrofoam in the chamber or down into the magazine well or stuck on the end of the bolt.

Buy the best optical scope you can afford along with good mounting bases and rings. Clean and assemble according to the manufacturer's recommendations. Once mounted, get a boresighter and do this procedure yourself. I have no faith in Big Box retailers doing this job for me.

Boresighting only gets you on paper and sometimes that does not work. That remains a mystery even to me. Set up your shooting table at the range with proper rests and seat using hearing and eye protection. Use only the same ammo that you intend to hunt with to sight in the rifle. Once the rifle-scope is sighted in, then for fun try other brands of ammo, but use the same bullet weights so you are comparing apples to apples.

Use the three-inches-high at 100 yards standard and work to get your three-shot groups printing at or under three inches. That is easy to remember. Let the barrel cool down between strings, especially if your rifle has a stainless barrel. I find stainless barrels tend to scatter bullets once they get too hot to handle. For that reason alone, I have no stainless hunting rifles. Lightweight, thin barrels tend to do this, too.

There is considerable debate about what scope setting to use during a sight-in session. I start with the lowest setting at first to get things locked in. Then I crank it up to full power to see if the groups change. They should not to any great degree. I have my scopes set to 6x nearly 90% of the time in the field. I rarely change the magnification, but that is just a personal thing.

Have I missed anything? Probably. Once you pick your next favorite hunting rifle or want to reconfirm an old setup, then follow these suggestions as a starting place to create your own lethal precision hunting rifle. Then hope that Ole Mister Buck Fever don't show up.

FOCUS ON OPTICS PLANET OPMOD BEST BUYS

In the world of hunting there can be no substitute for having quality optics at hand. I have found that out the hard way a number of times, but never again. So many quality optics are on the shelves these days that no self-respecting hunter of any game animal should be without a good set of binoculars and/or a high-quality spotting scope. Having access to both is the Holy Grail of big game spotting and observation.

The bottom line issue with good optics these days is that the prices seemingly have gone out of the roof and blurred the options for wide-ranging choices. With a tight economy still gripping our wallets and a considerable number of working class hunters having to stretch their recreational dollars even thinner, a high-quality product at a reasonable price quickly becomes a premium piece of hunting gear. Enter *Optic Planet's* own proprietary line of hunting optics marketed under the trade name of "OPMOD."

The OPMOD 8x42 binocular is an eight-power glass that is no wimp. These are fully waterproof with a textured gripping surface, which is a necessary feature for wet weather or snow conditions when hunting. The glass is fully multi-coated. The unit has a center focus wheel, a 42mm objective lens, 5.25 light transmission factor, and is nitrogen-filled with twist-up eyepieces. The binocular only weighs 1.5 pounds.

In field use these OPMOD binoculars are perfect for every outdoor activity from bird watching, football games, spying on neighbors, or counting points on antler racks at extended ranges.

The grip surface remains soft and pliable under cold and wet conditions, which means they do not slip out of the hands like slick surface optics tend to do. The sides of the optical barrels are dimpled with tiny gripping points like fine gun checkering. The center focus wheel is finished the same way.

The eyepiece focus is simple and accurate. I wear bifocals which are the bane of hunters trying to use riflescopes or binoculars. This is not an issue with these OPMOD binoculars. There are twist-up eyecups as well, but I found with my eyeglasses that they work ideally in the fully down position.

Out of the box these binoculars are packed in a nice, fully cushioned case with a nice, very comfortable neoprene neck strap. The objective lenses have soft material covers that can be popped open or removed altogether. The eyepieces also have a one-piece cover that loops into the carry neck strap. These are neat features that are usually only offered on really top end very expensive optics. This package retails for a bit over $200 but I recommend you check their website for often sales (www.opticsplanet.net) and promotions to get a better price. These binoculars are definitely a best buy considering the quality and utility.

Going up to more optical power via a spotting scope, the OPMOD 20-60x60mm Spotting Scope with tripod is just the trick. Oddly, as an experienced western big game hunter and now a fanatic white-tailed deer hunter, I never did seriously consider toting and using a spotting scope. That was a huge mistake.

Until I spent time in two different outfitted elk hunting camps out west years ago that convinced me otherwise. These adept outfitters used spotting scopes set up outside the camp house area for hunters to scan, observe, and spot elk way up on the mountain meadows and along forest tree lines. The observational advantages of using the spotting scopes quickly became apparent. These days I use my OPMOD Spotting Scope on a regular basis.

This OPMOD scope is a variable power unit going from 20x to 60x power with a 60mm objective lens. This offers a wide-open power range for wide viewing as well as up close and personal inspection of a big game animal and any rack he is carrying overhead. A spotting scope is perfect for really detailed evaluations of a set of antlers.

Specifications for this spotting scope include all the essential features needed for long-range game observations. The unit is water-resistant, lightweight at 2.2KG, offers a powerful zoom range, 4mm-1.3mm exit pupil distance, and includes a tripod which is ideal for a shooting range bench, working off a vehicle hood, or placing it atop a stand-up photography tripod.

The eyepiece is angled, making it easier to use in the field especially on a tripod since the user can simply bend over to view through the scope. Both lenses have soft rubberized protective covers, but they are not attached. The 20-60x power adjustment has a locking ring so the power setting can be dialed in and set fast. There is a large, easy-to-manipulate focus knob on the side of the scope barrel. The eyepiece lens also has a twist-out focus feature.

Another nice feature is the lock screw that allows the tripod mounting platform to be rotated around the circumference of the scope for different use applications. The exterior surfaces of the scope are fixed with a soft rubber-like material, offering a firm grip. The optical quality is exceptional.

Again, check their website for current pricing and further detailed information on the spotting scope as well as other Optics Planet optical offerings. Remember the "Best Buy" part.

QUALITY SCOPES DESERVE TOP PROTECTIVE COVER

One may ask just how much a product review writer could say about a simple device like a neoprene scope cover. Truth is, that is hardly the point. If you have a hunting rifle of any kind with a scope mounted on it, then you need a high-quality scope cover to protect it, and the *ScopeShield* is absolutely the best one I have used.

BASIC PREPPING ESSENTIALS: WEAPONS

And it isn't that this *ScopeShield* scope cover is just like any other optics protective cover either, because the unique design and features offer so much more in terms of user details with practical functions.

First and foremost, this cover is custom-designed in specific size ranges to fit the exact scope length you have. So, it is not just a universal cover sized M-L-XL to fit a general length of the scope tube and objective lens end. You have to measure your scope to get the right fit so it is a snug capture of the entire scope for complete protection.

Why a scope cover anyway? If you hunt much, obviously outdoors in the elements of dust, dirt, wind, rain, snow, sleet, muck, and mire of being in the woods or fields, then your scope is going to get abused just by being exposed to nature alone. Then comes the scratches, dents, and blemishes caused by getting in and out of vehicles, on and off all-terrain machines, climbing up and down tree stands, crossing over barbed-wire fences, and all the other rough abuse your gun and scope can encounter outdoors. It's part of the experience.

A high-quality scope cover like this one will help greatly deter most of the everyday bangs and scratches, but mostly it is just going to keep the optical lens more clean and clear of dust, dirt, and moisture that dots the glass. I never leave home or camp without one installed in place.

So what is so unique about the *ScopeShield* rifle scope covers? It is unique because of the "keeper loop" sewn into the cover body. This loop is used with an attached rifle sling to retain the cover when it is removed from the scope. The cover just falls free of the rifle and is captured on the rifle under the barrel by this loop with the sling in place. Without a sling it can be held by a rubber band, as well. This is a neat feature for sure and one I have never seen on any other scope cover.

Also the *ScopeShield* has a "grabber" loop on the eyepiece end of the cover that is used to release the cover from over the scope. This can easily be done even wearing heavy gloves, which is another feature unique to this scope cover. A small detail for sure, but once you use it, you will be happy it is there. There is no fumbling trying to get the cover off to make a quick shot.

To remove the cover from your rifle scope in the field, simply grab the "grabber" loop with the thumb and index finger of your trigger hand, tug it backwards and up, and the cover will easily fly free of the scope and be caught and held by the "keeper" loop. This is so simple it is almost silly.

Like I said, the *ScopeShield* rifle cover comes in many different size ranges and a variety of colors and graphics. I hope they make one in hunter orange soon. Check out all the product details, sizes, colors, and graphics available on their website at www.scopeshieldcover.com.

To keep your rifle scope lenses clear, clean, dry, and ready to sight that trophy big game animal, then install a *ScopeShield* on your rifle scope. Made in the U.S.A. in Oregon, this scope cover really is unique and the workmanship is top notch.

AIMPOINT'S PATROL RIFLE OPTIC

Today's marketplace is replete with choices for electronic red dot sights. It seems the primary objective for these sights is to place them on AR-15 type platform rifles, but certainly this is not the only use. If you have ever used such a sight, then you fully understand their utility. If you have not, then now is the time to try one.

Think of picking a good quality electronic red dot sight like picking the best can of green beans at the grocery store. How do you know which ones are best? In America, if you go to any food store you will likely see 10 or more choices just for a can of beans.

If you attend a big gun show with lots of dealers of gun accessories, you are likely to also see dozens of choices for red dot sights. So, shop well and buy carefully. There are some really fine electronic sights out there, and there are many that are pure junk. If you pay under $100 for an electronic red dot sight, what do you really think you will be getting? Yeah, right.

BASIC PREPPING ESSENTIALS: WEAPONS

Opposite from some of the grossly inferior red dot sights are products made by companies like Aimpoint, one of the most respected brands of electronic sights. Their sights have been proven by years of service on the battlefields around the world as well as countless law enforcement agencies working our cities and streets.

One of Aimpoint's "best" in my opinion is their *Patrol Rifle Optic or PRO*. One of the best features of this full-service electronic red dot sight is that is it designed to be easily used by shooters with both eyes open. Admittedly this takes a bit to get used to, but once you work with it, you will find it an exceedingly simple and effective sight.

The Aimpoint PRO is a serious-looking sight. From its hard anodized dark graphite grey matte finish, to its compact size, length of only five-inches and weight of 11.6 ounces including the integrated mount, the PRO is easy to set up and easy to use. Its dot brightness has 10 positions using one three-volt Lithium battery with three years of service on setting 7. Dot size is 2 MOA. This is an electronic red dot sight you have to try.

AMBIDEXTROUS FIREARM CONTROLS

While ambidextrous controls on various models of popular firearms are certainly not necessary, they can be a nice feature. They can also bail you out of a bad situation should your strong side arm or hand be taken out of play.

For the uninitiated, ambidextrous controls on firearms simply means mirroring such things as safety latches, magazine releases, bolt or slide closing selectors and such on both sides of the gun's frame or lower unit. These double-duty controls are most likely found on gun models like the 1911 semi-auto pistol and the AR-15 and clones, but are seen on other firearms as well.

The obvious overt use for duplicating control selectors on firearms is to make it more ergonomically simple for left-handed shooters, or for having to switch over from the right hand to left hand under certain circumstances. In terms of practical use, we can all train to use one set of control selectors typically found on the left side, but two sets has its uses, too.

Having field-tested many 1911s at the range over the years, I have come to appreciate in particular the dual safety selectors and wish all these pistols came with them. I find in steady shooting practice that I can even use my trigger finger on the right hand selector at times. Certainly shooting left-handed, which I highly recommend all shooters practice more often, the right-hand selector is essential. Otherwise you have to change hands altogether.

Many ARs shooters are now discovering the advantages of having a safety selector on the right side of the lower unit for the same reasons as having one on a 1911. However, the way the ARs are handled by most shooters, the trigger finger can easily be used to manipulate the right-hand-side safety switch. It seems that for other AR controls this is not so critical.

On factory production guns you will not see the dual selectors very often. The higher-end models like some Kimber 1911s will have them, but the basic models do not. This is true for AR-15s as well. For these rifles, as you shoulder them typically right-handed, then the primary control selectors are on the left side. There are plenty of aftermarket parts and kits now to upgrade just about any semi-auto firearm to add some dual controls if desired.

AMMO TOTES AND STORAGE BOXES

How do you carry your ammo? When you leave out for deer camp or the shooting range, how you pack and carry your ammo can make a difference. Sure, you can pile some boxes of rifle ammo into a cardboard box and throw it in the back of the truck. However, there are better ways to carry ammo to keep it safe and clean.

There are numerous brands of commercial molded boxes specifically intended for carrying ammunition. Two of the primary brand names for these boxes are *Plano and Case-Guard*. I am sure there are others on the market, too.

These boxes come in several different models, lid and latch types, so pick the ones that do the job best for you. In practice, the single hinge-lid tops seem to work well. I like the models with only one snap-over lock, not two smaller ones. The one-latch models seem to be tougher, thicker, and latch tighter.

There should be a way to add a padlock, slip tie, or even a cable tie if you want the seal to be more permanent for long-term storage. Also take note when inspecting these boxes for purchase to buy the ones with the rubberized gasket seal around the lip top. Some of these box lids are simple plastic to plastic. These could leak.

Take note of the handles on these ammo boxes, too. Some have larger, sturdy grab handles, while others are a bit flimsy. Once you load up one of these boxes, especially the smaller sizes, they are quite heavy. This requires a well-made, heavy-duty handle to tote the load.

As mentioned, these factory-made boxes do come in several sizes. I have noticed that some of them will not accept, for example, rifle ammo boxes turned a certain way. What happens then is you end up not getting as many boxes of ammo into the tote box as you might like.

I wish the makers would fashion these tote boxes more in tune with the more or less standard sizes of say .30-06 or .270 factory paperboard boxes. When you go shopping for an ammo tote, try out various sizes of boxes of ammunition to see how they fit or stack in the totes. I prefer the rifle ammo to sit on end, so I can read the ammo type on the end of the box.

Now, besides these smaller 30- to 50-caliber-type-sized ammo totes, there are bigger storage boxes available, too. While some of these can be used to haul large quantities of ammo from home to say a secondary bug out location, loading them up fully will make them very heavy. Still, most likely you will not be moving these around very often. They do take up considerable space.

These bigger boxes are intended more for long-term storage in a cool, dry, safe location. These boxes also come in two to three different sizes, so obviously different amounts and types of ammo can be stored back, locked up and secure.

If you have large amounts of ammo to store, then maybe one box for each type or caliber would be appropriate to reduce confusion. For example, one storage box could be used for only .223/5.56 ammo, or multiple types of hunting ammo, while another could be designated for just pistol ammo, or only 9mm and so forth. Add a strip of masking tape on the outside end of the box for a label that can be easily read when multiple boxes are stacked up.

Many of these storage-type boxes are quite heavy-duty and will withstand reasonable abuse. I carried one to Russia with me as checked baggage and even the six-foot, muscular Army lady could not destroy it, but she tried. These can usually be locked on both ends, too, to deter prying minds or outright theft, unless they carry off the entire box.

These boxes are ideal for secure storage for ammo. They do not leak as a rule, and will keep ammo dry, clean and safe. I would not recommend storing them loaded with ammo in a hot garage or up in an attic. A basement might be OK so long as there is not a high-moisture humidity level. But preppers need to keep in mind that these boxes can also be grabbed quickly for a bug out, so storing them on ground level is probably a smart move.

Keeping ammo together in one secure place is convenient for sure. Just make certain that access is limited and secured. If it is in a house closet, add a key lock to the door. Only adults should have access to such keys, ammo or guns anyway.

Using the appropriate ammo totes and storage boxes is better than just stacking random boxes of ammunition on a closet shelf or on the floor. These boxes promote organization, secure storage, and ready access when needed.

BULLET PERFORMANCE OFTEN A MYSTERY

What exactly do you expect a hunting rifle bullet to do? Truth is, actually, bullet performance on live game is pretty much a mystery in reality. This was confirmed one month while deer hunting at our camp. First of all, the conditions were pretty unusual for Mississippi. During the afternoon hunt it snowed for two hours with a blustery wind approaching 20-30 miles per hour. The snow blew horizontally.

That night the temperature got down to 26 degrees as the wind continued to blow. I was back in the hunting stand by 8:30 a.m. because historically deer have not been moving before about 9 a.m. By 8:45 a.m. I was tracking the blood trail of a big doe I had shot from the stand at 150 yards.

The doe did not fall at the shot. She bolted back into the woods in such a manner that I initially thought I missed the shot altogether. I was shooting a new rifle set-up and this was the first live shot at game. I sat there for a few minutes pondering my shot.

I was hunting with a new Remington 700 Tactical chambered in .308 Winchester. I mounted an excellent Leupold VX-2, 3x9x50 riflescope. It was targeted at the range at three-inches high at 100 yards. The rubberized Hogue stock allowed a firm grip and a comfortable recoil. I anticipated this set-up would make a great deer rifle.

The ammunition was nothing special, just a standard Remington yellow-green box of .308. The bullet used in this ammo was a 150-grain, pointed soft point. It was not a super duper bullet, but one that has been proven for decades as an accurate, game-getting round. It was designed to be ideal for deer hunting.

After sitting for a few minutes, I got down to check the spot where the deer stood. I immediately found spots of blood, so I began tracking. The blood trail was good and I found the deer not fifty yards from the trail.

At the skinning rack, I could see the evidence of the bullet performance. I shot the doe square front on in the chest. The bullet entered, then took a left hand turn and took out the entire shoulder, ruining all the meat on that side. The bullet did its job...too well, I guess. I am not complaining really, just passing on an observation. When a bullet hits real flesh, blood, and bone, there is no guarantee what it will really do, except its job.

CALDWELL'S SHOOTING BENCHREST LEAD-SLED

Some shooters are finding it helpful to use a rifle rest at the gun range. Among the very best benchrest shooting rests on the market are those marketed by Caldwell Shooting Supplies. There are a number of different models available to suit the needs of different shooters for a wide variety of shooting set-up.

The particular model I have field tested for over a year now is the *Lead Sled Fire Control FCX* version. The most notable features of the FCX model is the heavy base supplemented with a sand or lead shot bag up front to keep the rest from moving around. The bag itself is heavy duty and custom-made to fit into the front tray with a hand-carry strap.

Other features include an elevation adjustment knob on the side to raise or lower the platform supporting the rifle rest bags up on top. Once adjusted, having the rifle on target, there is another knob that locks the elevation position. A really neat feature unique to Caldwell rests is the control arm handle that can precisely move the resting rifle into the exact sighting point on the target.

This control arm handle takes some practice to get used to. The tension on the handle can be adjusted for fine movements as well to move the rifle's scope crosshairs up and down or right or left on the target. This allows for very precise alignment of the rifle-scope to the target.

The rifle sits in dog-eared sand bags on top of the adjustment platform as the buttstock sits into a contoured rear support. This support has a height adjustment foot on the bottom to raise the rear of the stock up or down. This support is attached to a rear frame

BASIC PREPPING ESSENTIALS: WEAPONS 93

support that can be slide adjusted to change the overall length of the rest.

I have used the Caldwell FCX Lead Sled now on numerous occasions to sight in hunting rifles, varmint rifles, and several ARs. Once boresighting is done and close range "on target" printing is accomplished at 25 yards, then work on the 100-yard range bench is greatly facilitated with the FCX rest.

Using the control arm handle to make minute adjustments of the crosshairs on the target cuts the precise sight-in effort down to nothing, saving time and ammo. The Lead Sled does not move around in the process either. You have to try a Caldwell Lead Sled.

OPTIONS FOR CARRYING BACKUP MAGAZINES

You may be a fancier of the tactical type vests with 50 pounds of gear hanging all over. I am not. If I were in the military, then I would be happy to carry all the loaded mags I could tote. This is not the general image I want to project even as a prepper or survivalist unless maybe I am in residency at my bug out location. I doubt I would wear one at home for a bug in, but I might if the action were on. I lean toward keeping a low profile.

Really, though, the point is how to comfortably and conveniently carry multiple loaded magazines and/or pistol magazines, AR-AK mags, or other ammo holders. I will let you decide which is more important, the convenient part or the comfortable part. Sometimes or usually one has to give into the other to remain expedient.

All you have to do is study the pages of a good supply catalog or web gear site selling ammo carry options and there are plenty of options to see. I have tried so many I could have my own garage sale with the webbing gear I have that I no longer use for multiple reasons.

Everybody is different in how they want to carry loaded magazines, but here are my primary modes of carry. On my person at a camp or bug out site, I use a nylon or heavy-duty leather belt separate from my pants belt to carry pistol mags in two or three mag pouches with fold-over secured tops. I use color-coded pouches for the three different type magazines that I might carry for one of the three different pistols I might carry. I rarely carry no more than 4-6 extra magazines on my person. I may have more in the truck or in a bag on the ATV.

Likewise for AR magazines, I use individual mag sleeves or three-mag sleeves that I put into a carry bag slung over the shoulder. I just don't like the vests, but you might, so try one. I also have sling-type multiple mag pouches that can carry up to six 30-round AR magazines. I may sling one of these over my shoulder and house extras in an ATV rack box or on the ATV rack. Search around for the best mag carry option for your modes of operation.

CHOOSING SHOTGUN CHOKE TUBES

In the younger days of my rural lifestyle I learned to shoot quail, dove, rabbits, and squirrels with a bolt action .410 Stevens shotgun, a hand-me-down from my brother. That diminutive little smoothbore actually had no choke so I would say it was an open choke.

Later I obtained use of another ne'er-do-well shotgun, an old Mossberg bolt gun with a drop-out magazine and an adjustable screw choke pressed on the muzzle of the gun. It was supposed to be screwed down to press or open slits in the muzzle end to control choke from improved, to modified, to full choke. It was worthless.

In those days most shotguns came from the factory with fixed chokes. So you bought a gun with a full choke if you wanted to hunt ducks, and generally a modified choke for small game. An improved cylinder was sometimes used for close-up shooting of busted quail coveys or sometimes running rabbits in open fields.

Later came the screw-in choke tubes which revolutionized the barrel choking of shotguns. I don't know who invented the screw-in

BASIC PREPPING ESSENTIALS: WEAPONS

interchangeable shotgun choke tubes, but my first Remington 11-87 shotgun had them, one improved cylinder, one modified and one full choke. Later I got turkey chokes and an extra full turkey choke for really tight patterns.

Essentially a shotgun choke tube provides the appropriate barrel restriction at the muzzle to pattern the shot in various widths or shot spreads downrange. An improved cylinder is a very open or wide choke useful for close quarters bird shooting. A medium or modified choke is designed for general use and small game hunting. A full choke is the narrowest or "tight" pattern for waterfowl hunting.

There are many other choke selection options, too, for buckshot or slugs, using an open bore choke (or no choke actually) or very, very tight chokes for turkey hunting. You can actually purchase other custom chokes for a variety of shot patterns for any number of specialized uses. Deer hunters can even buy special slug barrels for hunting.

Hunters can get away with traditional factory chokes for most applications. Preppers and survivalists using a shotgun for self-defense or protective uses will likely want an appropriate open choke for buckshot use.

Whatever choke is selected for hunting or self-defense, the shooter needs to pattern the shot used to see how it prints on paper to know its spread pattern. Check both lead and steel shot.

COLLECTING OLD GUNS IS COOL

Old guns are some of the best guns. Over the past few years, I have begun a transformation to my orientation back to my interest in old firearms. Sure, I like some of the newest and best models coming off the showroom floors. Ruger, Smith, Kimber, Savage, Glock, AR-Clones, Iver Johnson, Ithaca, and many others are bringing out some really great guns. Today's firearms are well-engineered and well-made. Alas, they simply do not have the allure of the old guns with their history and backlogs of service.

The old guns hold a real fondness in my heart. I encourage my readers to harken back to the guns of yesteryear, even if it is just for the study and knowledge renewal or for a first initiation. It isn't just one namesake brand either like the Colts, but old Smith and Wessons, Winchesters, Brownings, Springfields, Stevens, Rugers, Marlins, military surplus guns with eons of history behind them, and so many more. These are just flat-out cool guns to own and shoot.

I have others to blame for this ignition of interest in older guns once again, though admittedly I have been a student of the old revolvers and rifles all my life. Two of my favorite current gun writers are John Taffin and Mike "Duke" Venturino. These guys are profoundly engrossed in the old guns, so I refer to their works often. They do the grunt research, which cuts corners for me. They can be read regularly in *Guns* and *American Handgunner* magazines.

Likewise, I suspect they both blame folks like Elmer Keith, Skeeter Skelton, Jeff Cooper, and others whom I idolized as a young gun reader for their own affliction to the old Colt Single Actions, 1911s, Police Positives, Smith and Wesson Hand Ejectors, and so many more on a long list. While I favor the handguns, the old rifles and shotguns are another case of mental distress and suffering for the gun collector altogether.

I am often asked how to get started in this fevered addiction to gun collecting. First, buy some books even if it is a gun trader's manual or a book of gun values. Pick a couple volumes that offer photos of the guns so you can get an orientation to different firearms and the makers. Then perhaps buy a book exclusively on a brand that really interests you.

The tough part is narrowing all the choices down to what specific gun brands or even models that interest you most. That is not to say that you have to limit yourself at all. The two schools of thought are (though there is no law or rule of gun collecting that compels collectors one way or the other) to either try to collect one specific brand/model like the Colt Single Action Army, Browning A-5 shotguns, or

Winchester 1894s or such, or to choose collecting handguns, rifles, or shotguns exclusively.

Frankly, I prefer a smattering of it all, each choice being a representative example of a particular gun brand or model. I would be just as happy to own an original Winchester 1897 shotgun as a dozen of them. It would be very difficult and highly improbable to find samples of all the models and versions of any specific gun. That is if you could even afford to do it.

Once you make your choice(s), then you have to decide your level of quality or condition of the guns you are willing to accept to add to your collection. I tend toward shopping for as new as they come, 90 percent original condition or better. I prefer unaltered guns with clear name, model, serial number and proofs. If a screw head has been turned out or buggered up, then I pass. That is *my* standard. If the gun comes with the original factory box and papers, then all the better. Naturally they come priced accordingly as the gun books will educate you.

Then another real issue you will face is the book price and what I call the reality price. Gun value books today tend to be as much as 20-50 percent below cost estimates as to what you are likely to find on a gun tag at a gun dealer or gun show table. You have to plan and budget accordingly.

So, where do you find these old guns to collect? Anywhere and everywhere guns are sold. If you enter into this game, then you have to train yourself to keep your eyes and ears open constantly. Gun shops that take trade-ins or feature old guns will certainly have some. The stock can change daily, too. Many dealers selling new guns have quit buying old ones, but look nevertheless.

Bar none, the best place to look for a wide variety and quantity selection of old guns is at gun shows. The bigger the better, but shows over 500 tables can get overwhelming and unmanageable from a collector's standpoint. If you are up to it, take in the huge *Tulsa Gun Show* that gives you four days to look over some 10,000 tables. You'll need an FFL license from a cooperative dealer to buy guns out of state like that. And lots of cash.

At some shows you might find a dealer selling only Winchesters for example, but more than likely the collectible guns will be spread out potentially across every table at the show. You have to balance looking quickly to scan a table, but then be prepared for a long inspection if something of interest pops up. If you want that gun, then make the deal then and there. Otherwise if you delay to return, it may be gone. Let me count the times that has happened.

Take a gun value book with you and keep it in the car for reference. Have a bore light, too. Ask plenty of questions and don't be shy about dealing. That is all part of the game. Collecting old guns is cool. It is a neat hobby that never ends with new possibilities at every turn.

CONDITIONAL SHOOTING

One of the skills that concealed carry shooters and preppers alike need to practice is conditional shooting. What is that? Conditional shooting is simply taking advantage of any and all available nearby cover to protect one's self. If this sounds like pretty much simple common sense, then you would be right.

Hunters in the field and woods have long learned to camo themselves by using natural cover. If you get caught stuck out in the open in plain view, then you are simply busted by any game animal that spots you. If this were a domestic self-protection scenario, then standing out in the middle of a parking lot only invites a high likelihood of getting shot or otherwise harmed.

If you happen to ever be faced with a situation in which you must defend yourself in a public venue, then first learn to scan your surroundings. If you are inside of a vehicle, then you may need to lower your extremities to hide yourself so you can clandestinely observe what is going on without over-exposing yourself. This gives you ample time to prepare yourself.

If you are outside of your vehicle, in a parking area, in a shopping center, store or otherwise out in the semi-open, then immediately look around to see what defensive positions you could utilize to help secure yourself. This might be a light post in a parking area, another vehicle to hide behind, the corner of a building, a metal standing mailbox or something similar.

You not only want to reduce your own exposure to cover vital areas of your body, but also secure for yourself protective cover in the event you might be called upon or decide to engage yourself. At least you should be looking for cover from which you can view the unfolding situation. You may not decide to engage, but you still want to protect yourself.

Inside a store you may have to act quickly. I mean let's get real, you may never face a situation like a robbery or assault, but these happen every day and people are always caught off guard. That is what you need to plan and practice to avoid.

At the shooting range or information shooting practice, work on shooting skills from covered positions, kneeling, crouching, or other unusual shooting stances. Also practice with shooting offhand. Conditional shooting is a necessary skill to be learned.

DO BORESIGHTERS REALLY WORK?

You've seen them in the shooting stores or catalogs. There are several different kinds, now including models using a laser beam to project on the wall to align the scope crosshairs with the rifle's bore. But do they really work and just how precise are these devices?

I've seen this happen many times, but witnessed it again last week. I was in a mom and pop gun shop and watched the owner mount a new scope on a hunting rifle. Then he pulled out the bore-sighter and fiddled with getting the reticle on the scope to match the lines in the sighter. I would bet a hundred bucks the buyer of that scope never shot that rifle in at the range before going hunting. I wonder if he hit his target?

I have used a series of boresighter rigs many times over the past 40 or so years of mounting scopes and shooting rifles at the range. To me the whole process has really been a hit or miss proposition, no pun intended. Sometimes the boresighting is right on, other times I scratch my head wondering what happened.

Case in point was recently at the deer camp shooting range. I had a new Remington 700 Tactical rifle in .308 Winchester with a freshly mounted Leupold VX-2, 3x9x50mm, my all-time favorite scope. I performed the boresighting myself, or I thought I did.

I had two boxes of standard Remington Core-Lokt 150-grain hunting ammunition to sight the rifle in at 100 yards. I use the standard three-inches high at 100. For the first set of three shots twice, there were no holes in the target. None! I was dismayed, as you can imagine.

So, I did what I should have done first, and moved to 25 yards for a "print on the target" series. The holes immediately showed up in the lower quadrant of the paper target. Within six more rounds I was punching holes dead center on the target. So I backed up and started over.

By the end of the first box, I was hitting in the vicinity of where I wanted. Working into box No. 2, I got some respectable groups, three inches above the center, and into a two- to three -inch grouping. Not bad for old eyes using eyeglasses (which I hate with a passion). This rifle was ready to hunt.

But, what happened to the precision process of using the boresighter to "get on paper?" Honestly I have no clue. Had I gone hunting with the thought that the rifle was appropriately sighted in, I would have missed any deer I shot at. Also, being in a rush, I violated my own rule of an initial 25-yard sight-in just to ensure that the bullets were hitting somewhere on the target. Shame on me.

The lessons learned here are simple ones. A boresighter might well help get the scope and bore initially lined up within the ballpark, but nothing in terms of a guarantee on that should be assumed. This is where target range shooting at a known distance is essential to fine tuning the rifle for more precision shooting. This is also the part that many buyers of new rifles and scopes fail to follow through on. Hope you're not going grizzly hunting with that thing.

Also while I am at it, let me also point out a couple other critical aspects of mounting and preparing a new scope for its best performance. That dealer did not even have the shooter in the store to let him adjust the scope's mounting eye relief. He just mounted the scope in the center of the rings. That may work some of the time, but it is not a best practice.

When the buyer came in, he bragged on how great the new scope looked on his rifle. He picked it up (once), looked through the lens and pronounced it ready to kill a trophy. Oh, he also did nothing to focus the eyepiece to suit his own eyesight. I bet a good pair of wool socks it was not only blurry, but without a pre-set eye relief, there was a dark ring around the inside of the scope view. Hmmmm. Truth is, he might not have even known the difference.

So, when you mount a new scope, align for the proper eye relief to have a full, clear circle of light in the scope. Adjust the eyepiece to a clear focus, then lock it down with the lock ring. If the scope is a variable, double-check this with all power ranges on the scope.

Buy a quality boresighter and align the crosshairs. Then take the rifle to a set range and print a few holes at 25 yards to see what is happening. Get it centered there, then back off to 100 yards to finish the process to the sight-in height you want for the ammo you use.

I have never used one of the new laser boresighters, so I would be every interested in hearing from my readership on experiences using one of those. If they are as good as the press says, then I may have to break down and get one. Hey, Santa, did you hear that one?

FACTORY MAGAZINES ARE TRUSTWORTHY

Something makes me squeamish about aftermarket firearm magazines. Most shooters put full faith in their pistol- or rifle-factory-issued magazines at least and until there is a breakdown, failure to feed, function, or some other negative issue.

One has to assume that the OEM factory magazines are top quality and perform without issue or question. In my own practice, I have had very few magazines of any kind outright fail. Some needed an extra boost to engage them in the magazine well ready for action. Sometimes a magazine feeding lip will get bent or scarred, but even so that is rare for ones well-maintained and cared for.

If you drop yours a lot on the concrete floor at the range, or otherwise obviously abuse them, then your rates of failure or issues are likely to increase significantly. I recall once dropping a .308 magazine from my Rock River LAR-308 on a hog hunt that went down into the fine coastal sands of the Gulf Coast. That magazine locked up tight, and it took me forever to clean it enough to get back into service. Good thing I had a couple backups on that hunt.

So, what about pistol or rifle magazines not manufactured by the original gun maker? I suspect that trial and error is the best approach. When I shop magazines at gun shows, I see a lot of "no name" magazines for sale and I stay away from them. There are several brands of gun magazines on the market that have a good reputation for function and reliability. I would stick with those for sure.

In particular with the popularity of the AR rifles, the market seems to be flooded with all kinds of 10-20-30 and even 40-round 223/5.56 magazines. Some are aluminum, steel, plastics of various kinds, and some mystery materials. Again, stick with a known brand for quality like Magpul when picking an aftermarket magazine. Then, if a problem arises, the company will stand behind their products.

The ammo magazine is obviously a critical component of any semi-auto pistol or rifle. They have to be high-quality and function reliably every time the trigger is pulled. If you can't obtain an original factory magazine, then choose an aftermarket product with a good reputation that warranties their products.

CHAPTER 5
HANDGUN ACCESSORIES, GEAR AND TACTICS

CROSSBREED HOLSTERS: A BREED APART

More often than not these days, when I am walking my hunting property checking gates or perimeters or maybe hunting or scouting for deer or wild hogs, I am going to be carrying a sidearm. Sure, I will most likely have a favorite bolt action hunting rifle or a tactical AR hog rifle with me, but I also carry either a revolver or pistol for a variety of extra reasons.

Naturally then the concept of carry means a lot of things and the choices are too numerous to detail. Truth is, there are a lot of holsters out there on the market. Some are pretty good and worth buying, a lot of them are not.

I like simple holsters that are made for the gun to fit it right with proper tension to hold onto it and are also comfortable to wear. I mean, that is what a handgun holster is for, right? It is meant to carry your chosen sidearm at your waist or shoulder so it can be carried securely, safely, yet be drawn with confidence and ease when needed.

You've heard the cliché, "I've got a box full of holsters, but I'm still searching for just the right design to do what I want a holster to do." My search is over.

Simple, well-made, hand-crafted, durable, attractive, molded for the exact gun, well-fitting and comfortable on the waist. Back in 2005 when entrepreneur Mark Craighead tired of his overflowing box of holsters that did not work, he created Crossbreed Holsters. His holster designs are made for both inside the waist carry and outside the waistband carry for functional concealment or just regular carry.

In a true, full bore SHTF scenario I doubt anybody is going to be checking for a carry permit, but until that happens having the paperwork is a good idea. Then you can choose the option of carrying a weapon truly hidden by the legal definition of concealment, or just on the waist under an over-shirt or not. Crossbreed holsters work either way.

For my work and observation days afield, I just wear the holster out over my pants. I argue just having a visible method of defense could serve as a deterrent in and of itself. Maybe, anyway. I just prefer having that pistol butt where I can grab it.

To this end I picked out three models of the Outside Waist Band versions to test out. I have the Snapslide, Minislide, and the Superslide for the Beretta Storm Sub-Compact, the Smith and Wesson .380 Bodyguard, and a standard 1911 full-sized pistol. I have worn all three holsters both integrated with my pants belt and also with an additional over-belt. Either mode works, but frankly having a dedicated belt for the slide holster works best.

These holsters are made of a high-quality cowhide or horsehide back panel onto which is attached a specific molded Kydex holster sleeve. Each holster is handmade to precisely fit the exact model of the firearm the customer owns. These are not generic holsters or one size fits all types. They are not numbered for a general size range.

The Crossbreed models are slotted to fit a belt on the waist and can be positioned forward or back to make a quick reach comfortable for the end user. Some models have extra slots so a handgun cant can be adjustable on the waist.

In addition to the Outside the Waist Band models, Crossbreed also makes four models of the Inside the Waist Band models. They also make an Alternative Carry holster, magazine carriers, bedside firearm holders and a special line of high-quality belts. All models are suitable for Bug In or Out prepper plans.

A bonus for all Crossbreed customers is their policy of letting buyers use the holster for two weeks to fully check it out. If it doesn't satisfy the customer, then Crossbreed will buy it back. I doubt they get many returns.

If there is any doubt that a Crossbreed Holster can hang onto a pistol, forget it. As I used each holster I inserted the firearm, turned it upside down, and tried to shake the gun out. Obviously I did not toss it across the room or bounce it off the wall. Get real. Not once did the molded Kydex let go of the gun. However, when positioned on the waist in the proper manner, the firearm can be easily extracted from the holster using a firm grip and withdrawal.

So, next time when I am out in the Bug Out brush for hunting, scouting, observation or security, I will be carrying a sidearm in a Crossbreed Holster. I bet if you try one, you'll find its function suitable to your needs. Check out all the Crossbreed products on their website at www.crossbreedholsters.com.

MAGLULA LOADERS SKIP THE BUSTED THUMBS

Life can dish out its share of annoying little injuries. I hate paper cuts, mashed fingers in doors, a slipped screwdriver into the hand, anything heavy dropped on my toes, and thumbs busted trying to push rounds into a pistol magazine. Well, those days are over at least for the busted thumbs.

Loading brass cartridge ammo into single or double stack semi-auto pistol magazines can be a laborious, unpleasant chore. I don't know any shooter that gets his or her jollies out of jamming ammo down into a magazine. We all knew or hoped there had to be a better way. Well, there is.

Thanks to some brilliant engineering and product design work by Ran and Guy Tal of Maglula, Ltd., home-based in Israel, magazine-loading shooters now have a full line of Maglula loaders. *Maglula* is the acronym name that stands for "magazine loaders and unloaders accessories."

Their product line is extensive, so check out the website at www.maglula.com to see everything they make in loaders for rifles and pistols. They established their company in 2001 marketing loader tools to military customers and now commercial-private use markets.

Here I am focusing on the Maglula UpLULA loader for pistols in 9mm, .357 SIG, 10mm, .40, and the .45ACP. *UpLULA* is the Maglula product family name for their pistol magazine loaders, a personal Universal Pistol (hence, the *Up* designation) magazine loader.

I first encountered the UpLULA at my own hunting club range when hunting friend and shooter Gary Starkey took one out of his range bag and demonstrated its use for me. He was using a .40 caliber Smith and Wesson at the time with a double stack magazine and was loading them up faster than I could thumb press seven rounds into a Kimber 1911 mag.

Since I was using 10mm single stack magazines, he pulled out the alignment attachment accessory, snapping it on the bottom of the UpLULA loader to position the 1911 magazine perfectly for a quick charging of seven rounds. It was so slick and easy even I was impressed. I hardly wanted to give it back to him. As soon as I got back home to my computer, I ordered one for myself with the 1911 accessory piece.

The loader is just too simple to use. An empty magazine in any of the calibers listed above is inserted into the loader magazine channel box. The hand-squeeze grip is pressed in which engages an arm (they call it the "beak") that presses down on the magazine's follower.

When the follower is pressed down, a fresh loaded round is inserted into the magazine with the primer to the rear. Repeat the action until the magazine is loaded. It takes less time to fill a magazine than it does to describe it. I am certain, too, that there are YouTube videos on-line to guide you in the loading process as well if needed. One instructional video should do the trick. It really is that easy to do.

If you have straight body single stack magazines like the standard 1911s then you snap on the accessory part on the bottom of the loader and proceed as above.

The Maglula UpLULA loader is made of a tough, heavy-duty plastic product. The process of producing their loaders is highly detailed, including prototypes made by hand from metal, plastic, and wood, then precise 3D CAD models are created leading to injection molding processes to yield the final products. The loader is lightweight and easily fits into a range bag or even a shirt or coat pocket.

Quality control inspections are rigorous. This includes individually inspecting all products prior to packaging and shipment. Random samples are inspected and thoroughly tested as well. Just knowing their manufacturing processes can give the shooter a high level of confidence that they are buying a top notch product capable of complete reliability and long-term functionality.

Maglula also makes a *Baby UpLULA* that is a smaller version of the one described here. The "baby" model functions in the same manner to reload pistol magazines for the .22LR through the .380 ACP. That includes the .32 ACP as well. Of course, they make an extended line of loaders for the .223 (5.56) and the .308 (7.62). Again, check out their website for complete details and product descriptions.

So, there is no need to wear out your thumb and hand pressing down cartridges into a pistol mag one at a time. With the Maglula UpLULA you can quickly reload empty magazines and get back to the fun part, shooting the pistol.

GUM CREEK'S CONCEALMENT HANDGUN HOLSTER MOUNT

Concealed carry is very fashionable these days. Across the country the number of individuals taking basic and advanced concealed weapons courses is on the steady increase. With most advanced courses requiring a training and shooting session at the firing range, more people are taking concealed carry seriously.

The primary reason is for personal security on the person or for carry in a vehicle when traveling. Outdoors people on the move to recreational venues, to hunting camps or the local lake for fishing more and more feel the need for an armed backup. It's a different world we live in now. Even highway rest stops have armed guards patrolling parking areas now and other stops can be equally as dangerous to the traveler.

One of the questions that often then arises is what is the best manner to carry a handgun in a vehicle that is both readily available and also concealable to any outside prying eyes? If the vehicle driver does not wish to be wearing the gun on their person inside the vehicle, then the choices become the available reach and grab options inside the car or truck.

The options can be limited in most cars, SUVs, or pickup trucks. Center consoles either have a top that closes down or the weapon would be sitting out in the open for it to be reached quickly when needed. Glove compartments are completely out of the picture as the reach (especially if the vehicle is still moving) is cumbersome, dangerous, or even beyond reasonable reach. Some latches can be problematic to open quickly as well.

Likewise, the side door panels might be a choice, but can be difficult to reach into to extract a handgun while concentrating vision through the windshield. Even a pistol stuck down between the seat and the console is far from a best choice, as would be simply putting the gun down on the floorboard in front of the car seat.

The overhead visor would be a poor choice as well. There would be difficulty in securing a pistol above the visor and holding it there without it slipping out. And what if you actually needed to use the visor as a sun visor? Then the handgun would be in full view or it might fall out into the driver's lap. Let's hope the safety is on, or the chamber is empty.

Then speaking of the best type of weapon to carry in a vehicle could be a topic of campfire discussion that would potentially last all night. Wheel guns are a great handgun weapon in the hands of a trained user, but they can be bulkier, thicker, and harder to conceal. Pistols are generally flatter, but more complicated to use, especially considering the various safety mechanisms that pistols deploy. Training on any chosen concealment weapon is paramount.

Then along comes a forward-thinking company by the name of Gum Creek Customs with their *Vehicle-Handgun-Mount* line of vehicle gun concealment holsters and mounts. Simply put for the layman to understand, these handgun mounts go under the steering column of your vehicle where the gun can be easily reached with your right or left hand, but remain well-concealed under the steering wheel mechanism. Even a casual passerby would likely never see a handgun mounted in one of the Gum Creek mounts with the proper holster in use.

"Quick and accessible, our universal steering wheel column mount is ideal for carrying a handgun in your vehicle for convenience or self-defense. Great for those who often conceal carry and may not want to keep their pistol on them while driving. Also popular for those who travel or as a self-defense system for situations where a weapon is desired to be in immediate close reach."

The Gum Creek mount hooks go directly into the gap below the vehicle steering wheel column and secure at the bottom of the dashboard or lower kick panel. The mount strap is then pulled tight to draw the holster close up to the panel below the steering column. The driver/user can easily release the mount system in moments to securely put it away, take it out of the vehicle altogether, or to put it in a backpack or briefcase when departing the vehicle.

Many standard holster models can be used to attach to the Gum Creek mount or a holster specific to your firearm can be purchased from the company. Holster sizes and options can be seen on the company website at www.gumcreekcustoms.com. Optional mounting straps are also available for vehicle models that don't have a mounting gap below the steering column.

So, if you want to carry a pistol with you in a vehicle that is both secured and concealed, the Gum Creek Custom rig is the way to go. It will be easy to reach, but remain out of sight from strangers casually passing by the vehicle.

ARE CCW GUNS POINT AND SHOOT ONLY?

Now that's a good question. Using a CCW firearm can run the gamut from heated, emergency short-range shooting affairs to the possibility of longer range shots within realms of reason. These are certainly some parameters worthy of considering if we are going to carry a concealed weapon. Truth is, we never really can know under what precise circumstances we might be called upon to draw our CCWs.

Typically, we think of a "CCW" moment as a close quarters assault on our person, family or friends, perhaps out in public, or maybe as we drive into our own home driveway or garage. We could be out in a public parking lot outside the grocery store and be approached by a knife-wielding thug intent on robbing us or worse.

Usually we frame such preparation for any such event as a one-on-one situation. That may or may not be valid. There is also the realm of possibility that some may elect to intervene in a situation or threat that does not direct involve them. For example, witnessing a robbery, an assault on another person, or somebody shooting others such as in a movie theater or school environment. There could be more than one assailant, multiple guns ,and at different ranges with varying obstacles. See, it is never as easy as we might think.

Having just presented a few possible scenarios then, again we most likely think of a CCW offensive or defensive move as being short-range, perhaps under 10 feet. Most one on one weapons self-defense exchanges are from three to seven feet. Even so, we may be called upon to make longer shots.

So, are we better off with some high-profile adjustable sights, low fixed sights, night sights, Tritium sights, red/green glow sights or perhaps even a smooth no-sight pistol slide? Should we practice to shoot rather instinctively over the barrel (slide) or actually focus on sights? Or go one technological step further with a laser grip pointer, or a laser mounted ahead of the trigger guard, or a laser built into the body of the pistol like a Taurus Curve or a Smith and Wesson Bodyguard? The options are many.

I doubt seriously a fully adjustable high-profile sight would be needed on a CCW gun. The best choice is either a low-profile open-fixed sight and/or a laser sight integrated into the shooting plane. You want the CCW gun to draw fast and smooth and point naturally.

ARE HANDGUNS A WEAK LINK?

For all intents and purposes, handguns are the top-selling firearms in America right now. There are a lot of practical reasons behind this handgun use movement, but the truth is that a handheld firearm may not always be the best choice for every self-defense situation.

Also, concealed carry is a huge paradigm in this country right now. State after states are passing laws to allow citizens with the proper credentials and approved background checks to get a concealed carry permit supported by the completion of proper training. This means handguns, too.

However, a compelling piece of research data has indicated that even police and other law enforcement personnel prefer a shotgun first, and an appropriate rifle second over a pistol for self-defense and armed encounters. That fact alone should cause a pause for thought.

Handguns are well, held in the hand. This means they can be more difficult to grip, control, aim, and fire with consistent accuracy. The line of sight down handgun barrels from 2-8 inches is considerably shorter than the sighting plane down a rifle or shotgun barrel. Hence, a handgun is more difficult to line up on a target and harder to keep it there.

This demands considerably more practice and regular trips to a shooting range or the back 40 to keep proficient with a handgun. Of course, shooting practice is needed too with rifles and shotguns.

Furthermore, obviously handguns are hampered by their lack of relative power and the range they are useful for in active shooting situations. Most confrontational self-defense shooting incident ranges are seven feet to 10 yards. Forget seriously trying to engage a target at 50 yards out or more with a handgun. Sure, you might keep their heads down, but you're not likely to hit them or incapacitate them with a handgun caliber hit.

A really serious threat can be quickly thwarted with a shotgun blast of buckshot out to 20-30 yards. A rifle such as an AR-15 can engage targets with effectiveness out to 200 yards or more depending on the experience of the shooter. A "sniper"-type rifle in the right hands can deter threats out to 1,000 yards.

Now, I am the first to say, sure you have to have a handgun, but only if you know how to use it well, know the controls, and can manage it in action. Just remember, the handgun is not the end-all weapon.

ATTRIBUTES OF KYDEX HOLSTERS

Whoever invented Kydex® material for use in firearm holsters got it right. For those shooters interested in having a concealed weapon carry on their person, the issue always comes up about the best way to holster a CCW handgun. Holsters made of the Kydex material is among the best options available.

Kydex was developed in 1965 as a thermoplastic material originally designed to be molded into aircraft interiors. If you fly much today, just look around the cabin of every airliner and you will see the molded material around the windows and elsewhere. This is Kydex.

Today, the material Kydex is owned and produced by Kydex, LLC for a multitude of uses including the base material for many types, sizes, and configurations of handgun holsters and knife sheaths. For these purposes, this plastic material is near ideal.

Kydex thermoplastics are an extremely flexible material for manufacturing purposes to be used for holsters. The plastic sheet materials can be produced in any color, various thicknesses, textures, and can be heat press-molded into virtually any shape. This is probably why shooters see so many holster products on the market today made of this material.

The material is so easy to work with that I have seen a vendor at local gun shows making Kydex holsters customized on the spot to fit any handgun a customer might have. He also keeps many dummy pistol models on hand that he can heat-mold with the machine he has right on the vendor table at the show.

He makes holsters in various colors, material thicknesses, and some with customized safety snap-on straps and other accessories. He makes IWB and OWB holsters as well as traditional belt loop holsters to wear on a pants belt or secondary waist belt.

Kydex makes a good material for a concealed weapons carry holster because a pistol fits the holster tight and securely. The material is rigid but also has flexibility to bend and twist with body movement, sitting, stooping and bending over. The material is impervious to sweat or moisture and will not change shape or deteriorate like leather or even nylon.

Some do report that having a concealed holster made of this material against the skin is hot, but then I find any holster next to my body gets hot. I don't know of a solution to that. It is more important to have a secure fit with an easy withdrawal of your weapon.

CONCEALED CARRY OPTIONS

Sometimes too many choices can make things confusing. Thus seems to be the case with picking just the right concealed weapon holster carry. Carrying a concealed weapon on your person is serious business. Most people probably hope they never have to draw a pistol out in self-defense, but yet want to be able to do so effectively if the dire situation should arise.

First and foremost is the balance between conceal ability and comfort. Naturally you want a concealed gun, well-concealed. This means completely hidden from view by another person walking down the street, across the room at a meeting, or in any public place like a grocery store, restaurant, or department store.

Concealing a pistol on your person can be done in many different fashions, no pun intended. The deal though is that you have to try out the various modes available and pick one that universally works for you. Narrowing the choices can take some time and research.

Then you also want the piece to be comfortable to carry especially if it is typically going to be an all-day affair. You don't want a concealed gun and holster snagging on everything you walk by under your shirt, or rubbing an abrasion against your skin inside a waistband. You also want a pistol pouch that is easy to access smoothly and quickly.

This is assuming for example that most men will want to carry theirs on the waist or in a pocket. Women have other options with special purses and bags designed with slip-end pockets to insert a concealed pistol. But what if that purse is out of reach just when you need it? Things to consider. Some women might want to consider a pants pocket slip holster.

So, folks, if you belt the pistol there are several options. There is "inside the waist band" or IWB or "outside the waist band" or OWB, or in the pants pocket, or inside the pocket "holster sleeve." IWB or OWB holsters often clip on a pants' belt from the outside or the metal/plastic clip slips over the waistband from the inside. The holster accessory market is jammed with choices of all sizes, shapes and materials.

Whatever type of concealed pistol holster you choose, make sure that the pistol fits it well and won't shake loose. Sometimes concealment takes precedence over comfort, but it is a close call.

DO CROSSDRAW HOLSTERS WORK FOR CONCEALED CARRY?

Concealed carry modes seem to be concentrated on IWB or OWB carry formats. I don't want any pistol carried inside my pants and am not too crazy about many of the outside the waistband holsters either. Though I know many in law enforcement or other security positons that require such carry, I note that many of them complain about the comfort factors of carrying concealed.

If you have ever tried to carry a pistol all day in a holster fitted inside the belt of your pants then you know what I mean. Then try to sit in a car seat or at a work desk with this same setup. While standing or walking these carry methods do not seem to be too bad, but sitting is another matter.

To my way of thinking, too, even carrying a concealed weapon outside the pants belt has its comfort issues, too. When I do carry, I use one of the slip in the pocket pistol sleeve "holsters." It is not ideal either as the weight and bulk of even a small pistol wears on you after a while.

It occurred to me recently that maybe I should revisit the crossdraw carry for concealed carry. But even having mentioned that type of handgun position as concealed is problematic and I can hear the dissenters now.

I fell in love long ago with the crossdraw carry when using handguns for deer hunting. In practice afield the crossdraw carry is the most comfortable mode of carry I have found. Whether sitting in a vehicle, on an ATV, or in a tree stand, the crossdraw presents the easiest way to draw the weapon as any holster configuration there is.

I have Model 111 Bianchi crossdraw belt holsters for every hunting revolver I use. I also have a large nylon chest carry crossdraw for my scoped Redhawk and a Smith 460 hunting handgun. Then I have had two custom-built Diamond D Custom Leather Guide's Choice chest crossdraw holsters made up in Wasilla, Alaska. These are shoulder rigs that position the holster right in front over the chest for super quick access.

Now I know these chest carry rigs require the use of a coat for any type of concealment, but the belt carry crossdraws are easier to conceal. I'd like to hear your thoughts on this!

THE DESANTIS MINI-SCABBARD HOLSTER

DeSantis holsters are the Chevrolet and Ford of handgun holsters. You might throw in a little Cadillac and Lexus as well. What I mean by those comparisons are simple. Chevy and Ford are the workhorses of the automobiles and trucks in America. They work hard, get the job done, and last forever if maintained well. Then the other two brands exhibit the flash and beauty of a product. DeSantis holsters embody all of these attributes.

Since I acquired a SCCY, 9mm pistol as a Christmas present from Vernon and Helen Graves of Renegade Sporting Goods for helping them work gun shows, I have been searching for just the right carry holster for it. I did not want a common nylon cheapie, or any type of a custom-made holster that came with a payment plan. I wanted a good, solid leather holster I could carry on a belt. I shopped long and hard, because I was in no rush.

Then a magazine ad caught my attention and I began to study the DeSantis Gunhide web site at www.desantisholster.com. I followed that up with a visit to my bud's gun shop. He carries DeSantis holsters but did not have a model in stock to accommodate the new(ish) SCCY pistol. So, I got on-line to look further then order one.

I selected their Mini-Scabbard model, which I had seen and inspected at the gun shop. DeSantis makes this holster for virtually every small pocket-type pistol produced including the SCCY pistols. This holster is high-quality leather, but without too many bells and whistles. It can handle belts up to 1-½ inches wide and comes in black or tan, unlined. It has a tension adjustment screw to flex the pistol fit.

So far in my practical carry of this holster, I know I made the right choice. It rides in just the right spot for me and the holster cant puts the grip of the SCCY in an ideal position for quick access. The holster rides great on a pants belt without undue weight or sag. Using a second belt, I can slide the holster back off my hip as well. Under a shirt or jacket it can easily be concealed for CCW carry. Retailing for $61.99, I think a lot of shooters will want to check out the DeSantis Mini-Scabbard.

THE DESANTIS THUMB BREAK SCABBORD FOR LARGER PISTOL CARRY

For many concealed carry proponents and individuals wanting to carry a substantial gun for personal protection, toting the classic 1911 pistol can be problematic. Despite really big guns like, say, a Desert Eagle in .50 cal, a standard 1911 Colt .45 ACP or one of the hundred clones weighs in at around 38 ounces unloaded. That's a pretty hefty waist carry gun, but the power it packs is well worth the effort.

That's 2.375 pounds lugging on the waist. If you intend to carry a pistol the size of a 1911, even a short slide model, you will need a special holster for concealment. Such a holster has to keep the pistol tight next to the body, but in a ready draw position. Ideally this holster would create as minimal an imprint as possible to keep the carry discreet even under an over-shirt or jacket.

The candidate I have in mind for this setup is the DeSantis Thumb Break Scabbard. The minute I slipped a full-sized, five-inch slide 1911 into this holster, I had the idea right away this one was going to be perfect for carry under a classic Hawaiian-type shirt that is worn outside the pants.

In use I have also found it perfect for carry on the pants belt or an over belt rig while riding an ATV around my bug out camp just checking on things. Easily reaching down to grab the grip is essential in a carry like this, but the DeSantis Thumb Break Scabbard rides high thus positioning the pistol grip at just the right forward cant for a quick, yet firm retrieval.

With very little practice the reach becomes virtually second nature, as it should. You certainly don't want to be fooling around with drawing your weapon when time is of the essence. As with deploying any weapon, I recommend practice in all aspects of using it, so as with any holster, be sure it rides right where you want it and your draws are comfortable and flawless.

In this holster's case, too, the thumb break feature adds that extra element of security to weapon retention. The thumb break holding "flap" is held by a secure snap. Some models also have adjustable tension. Available in black or tan, smooth or weave, it retails from $76.95 to $107.95 depending on gun model, at www.desantisholster.com.

FAVE HOLSTER POSITIONS

A lot of discussion is devoted in the self-defense arena about concealed carry and what gun or holster to use, but little to nothing about where to position that weapon. Are there best carry positions to use? Are some modes more accessible than others? Is there a more comfortable way to carry a handgun concealed or otherwise? Good questions.

If we limit this for now to concealed carry, then we are talking about carrying the weapon inside the waistband or outside of it. This the carrier has to decide as a matter of personal preference. Inside the waistband means of course inside your pants with the belt lapping over the outside of the holster. I do not like this mode simply because it is not comfortable for my body style, which is no style at all. Try IWB to see if it works for you.

Outside the waist means on a belt either with the belt fed through the holster loop, or a clip-on style that secures over the pants belt. Then a shirt worn outside the pants covers the weapon, and/or a jacket provides the concealment. With either approach the covering garment has to be swept aside to make the weapon withdrawal. This takes practice for efficiency to get the weapon drawn and aimed in a timely manner when reacting to a quick response.

The next consideration is where exactly on your waist in or out to place the holstered weapon. The majority of concealed carry modes generally just position the holster directly at the side of the waist just under your drawing hand either right or left side. Others move the holster somewhat further back just behind the hip bone so it is somewhat obscured by the hip.

Another choice is to position the holster in the small of the back, which tends to be just to either side of the spinal cord and not directly over it. This can provide an extra measure of concealment but it is not particularly comfortable for sitting at a desk or driving a vehicle.

Other options include crossdraw which is really good for sitting but requires a further reach. Shoulder holster carry is an alternative if you regularly wear a suit coat or jacket. So, try out various modes to see what works best for you, then practice drawing the weapon.

GET A GRIP

What criterion do you use to purchase a new handgun, especially a concealed weapon? Is it by overall size of the firearm, the caliber, the barrel length, magazine capacity or what? Do you have a ranking system you use to make the best final decision for the "right" gun to buy?

How about considering the grip? Yeah, that handle thingy you grab ahold of when you pick it up.

I am amazed at how many new guns I read about that I would like to own until I pick it up. Admittedly I have big mitts, but most of the smaller, compact, concealed guns just do not fit my hands well. In fact, they don't fit my shooting hand well enough to make me want to buy it. I mean, if I cannot get a good, firm, hand-filling hold on a handgun, then how can I expect to shoot it well? I don't think I can.

I'm not picking on certain models, as some guns are just small guns, and that is that. Even so, I find some tiny guns fit my hand better than others primarily because of the angle of the grip and frame. A good example, is the diminutive Beretta 21, a small, .22 rimfire semi-auto pistol. When I grab up this fine little pistol, it just fits, thus I am able to shoot it well.

Another small pistol that fits me well since it is styled after the class 1911 format, only downsized, is the Kimber Solo. The grip fits right, the grip panels are comfortable, and I can reach the safety and slide release controls from the grip with ease.

By contrast, I really like the whole new line of Sig-Sauer compact, concealment pistols, but I hate them. I'm sorry, but if I have two fingers falling off the bottom of the grip then this is not a gun I can shoot well.

So, be sure after you pick a brand, a caliber, and a gun size, that you examine the grip fit seriously. Your grip hand should just naturally fit the pistol. It should be comfortable, not too small, too large, or too thick. The factory grip panels should be comfortable, too, or changeable.

When you point the handgun downrange, does the grip remain stable and secure? If your handgun's grip does not fit or feel right, then maybe you need to search for another handgun option.

HANDGUN GRIP MATERIALS

Handgun grips are for, well, gripping. Some are better than others at accomplishing this task, so gun buyers should be aware of how a handgun fits your shooting hand, but also take note of the type of material they are made of for a number of reasons.

Since handheld weapons have really come into popular vogue, the gripping designs have changed radically. Early black powder handguns basically had stocks that included short forearms to support the barrel, but the hand grip and stock was all one piece of wood back in those days.

Over time, as handgun designs evolved into metal gun frames, the manufacturers began putting on externally attached grips or panels. These were made from a wide variety of materials though mostly they were made of wood or a plastic-synthetic material of many different types of compositions. Aftermarket handgun grips are also available in soft rubberized materials for recoil absorption such as those made by *Pachmayr*.

Handgun grips can have many different kinds of textures, too. Some of these such as smooth wood permit the handgun to recoil in the hand without any particular abrasion impact. Sharply checkered grips allow for a firm grip, but can be rough on the palm of the hands, especially in heavy recoiling handguns in .357 Magnum, .44 Magnum, or some of the pistol calibers such as .40 S&W. When you buy a handgun be sure you can tolerate the type of grip texture on the gun.

Modern synthetic materials for handgun grips are very popular now. Top on the list are grips molded from various kinds of acrylics. These can come in an endless array of colors or molded in designs, logos, tradenames, or other features.

Also highly popular today are grips made of the super tough G-10 epoxy fiberglass. These grips can be milled down to an exact shape to fit practically any handgun model. G-10 grip panels are often found on upscale models of 1911 pistols and come in smooth or checkered versions in a wide variety of colors and combinations. G-10 grips are extremely durable.

Wood grips are still very useful and practical. The aftermarket is flooded with brands and varieties. Some are laminated types, bringing a mix of colors in stripes or other configurations. Popular now are grips that combine wood or laminated wood with soft rubber inserts to absorb recoil. So, when you buy a handgun, also know the type of grips you are buying.

HANDGUN HOLSTER CHOICES

Holsters are designed to carry a handgun. Unfortunately it is not as simple as that. Most shooters of all kinds of handguns, revolvers, or pistols probably own several boxes of various brands and types of holsters in a host of configurations and materials. Some are still searching for that one perfect holster for a favorite carry gun, but I suspect it may not exist.

First, handgun carriers have to decide why they are carrying and how they want to do it. Will the gun be concealed or open carry? Will it be a strong side draw, grip forward, grip back, cross draw, off hand, inside the waistband, outside the waistband or even a shoulder carry? See, there are lots of options. The primary holster you use should be mission specific.

Certainly users may chose a variety of ways to carry depending on the mission or the intent of the day, weather, temperature, audience, etc. In theory you might have two to three or more holsters for the same handgun to give yourself a variety of carry options. There is certainly nothing wrong with that.

Buying a holster for a specific gun begins with picking a brand/model that is supposedly designed for your gun model to fit. This needs to be verified by inserting your gun into the holster, then trying out the carry around the store before buying it.

Many holsters are designed to adjust the tension on the handgun by tightening or loosening a screw adjustment built into the holster. Other models just offer one fit and your hope is that the holster holds your weapon without releasing it unduly while in action. As you fit your gun into a holster be comfortable and secure in its fit. Make sure you can extract the handgun with relative ease without undue drag. Turn it upside down. If your gun falls out, then pass.

Holsters today come in a variety of materials, the majority being leather, nylon, or thermoplastic. Each has its own advantages and downsides. The thermoplastic does not wear out and seldom breaks. Nylon holsters can wear and leather stretches over time. Any holster has to be maintained even if it is just wiped down with a clean, damp cloth. Leather should be treated with proper care for longevity. Get the proper mode for carry and the right fit for your gun.

HANDGUN OPTIC IRON SIGHTS VS. BLACK SIGHTS

Trends come and go in the world of firearms development. Way back in the day, a six-gun had a groove cut down the center of the top strap of the frame and a half-moon sort of front sight blade. That front sight could be filed down to change the gun's shooting point of impact. Pretty crude for sure by today's standards.

As Loretta Lynn's song says, "We've come a long way, baby." Iron sights on handguns, both revolvers and pistols, have undergone a tremendous course of development and refinement since those early handheld weapons. Some would say in some regards the trends have reversed effectiveness, but such is in the eyes of the beholder and the user.

I never will forget the first Smith and Wesson revolver I bought with what I considered at the time truly fine fully adjustable rear sights and an excellent trademark front ramp with the red-orange insert. Those iron sights were the sights to end all sights. Well, hardly.

Today, what is in vogue are open iron sights on handguns with modern optical enhancements. The far extreme of these are the day-glow inserts put into holding mounts on the front end of the barrel. Some of these glow sticks are red but others are bright florescent green. Indeed, these newfangled sights are easy to see and pick up in low light conditions. They work for that.

There is some reasonable concern about the sturdiness of these plastic inserts. While I have not heard of these breaking or falling out, the possibility is certainly there. Users of guns with these sights will have to determine their true value.

Regular black iron handgun sights have served well for decades. There is something unique in the engineering that just works. The thinner front sight is positioned into the rear sight half-square leaving some light around the edges when the front is centered. Some rear sights are also fixed with a white perimeter around the open sight for further help in front sight alignment. These can be fixed or adjustable sights.

I have held up handguns with the glow sight inserts in gun shops and gun shows and they definitely do gather the light. One assumes they would perform the same out in the field. My only initial concern is how fragile they would be in active use. Time will tell on that one. Try them out for yourself and see what you think.

ASSUME THE HIGH READY POSITION

Armed self-protection is a big deal now. And for darn good reason, too. It is a scary world out there and getting scarier by the minute. This means of course a sharp increase too in concealed weapons permits, enhanced permits with the range shooting requirements, and a lot more classroom training ranges popping up all over the landscape to handle the shooter demand.

I had my first epiphany several years ago. I was in an off-downtown part of the city in a high traffic area at a gas station just off a main travel artery. As I was pulling the gas pump handle out of the tank cap of my vehicle and turned around to hook it back up to the pump, I bumped into a disheveled man that had crept up behind me. I mean nearly touching me.

I instinctively stuck out my arms, shoved the guy away and chewed him out for coming up on somebody like that. Later it occurred to me, what if he'd had a knife or gun and was intent on robbing me? I was not prepared. And even if I'd had a concealed gun would I have been ready to effect the *High Ready* shooting position needed for really close quarters encounters?

Laughingly I describe the High Ready as shooting from the armpit, but viewing this hold position it appears much like that. We are talking about executing a pistol shot from a range of one to three feet away, maybe even somebody right up against you.

The *High Ready* is a close-range technique for shooting your pistol held high up against your body with the pistol right against your upper chest side. You shooting arm is bent at the elbow to pull the gun in tight. The gun needs to be held about a full hand's width from your chin or cheek. It has to be held down below your peripheral line of sight, so you can see around you.

The hard part is practicing this hold and shooting position while trying to keep the gun's muzzle square horizontal with the ground. It has to be shot straight from this retracted position. This takes a lot of practice and getting used to a gun going off too close to your ears, eyes, and face.

Try the High Ready position by dry firing, then, when ready, post up a human form target and try some shots from near point blank range. It takes some getting used to, but might save your life.

HOLSTER POSITION CARRY OPTIONS

So, you have your handgun ready to carry. Maybe this is your first pistol or revolver and right now you are unsure exactly how you want to carry it or if you intend to carry it concealed. What are all your options for carrying a handgun on your waist belt?

First, make these decisions. Will you carry concealed with the gun totally hidden from overt view or open carry in plain view? Obviously you need to thoroughly check the carry laws in your state to see what you are permitted to do. Messing up on this one could result in some serious trouble, including a felony in some states. Make sure you are right about your carry mode.

If you go concealed, then your options are inside the waist band or IWB or outside the waist band or OWB carry. You will see many models of holsters so designated so be sure you know exactly what you are buying and if that is the right choice for you. Ideally you can test the holster fit on your body before you commit to one.

IWB holsters are certainly easier to hide, but many of them are less comfortable to wear, especially for long periods of time. OWB holsters need to be covered by a shirt worn out or a jacket. Sometimes the OWB holsters can slip around the waist, but if on the pants belt it will be contained by the belt loops.

Study the cant of the holster so that a quick, easy, smooth draw is facilitated. This is different for everyone. Some like their pistol grip canted forward for a quick grab, while others like the grip up close tight to the body in an upright position. Try different modes so you know what feels and works best for you and the gun (size) you carry.

Another option to consider is a crossdraw carry. This mode makes for an easy draw if seated say in a car seat. The movement of your draw hand across your body does take longer, but we're only talking miliseconds. For prepper work, the crossdraw method of carry is great for riding horses or ATVs for bug out work or patrolling property on security rounds.

BASIC PREPPING ESSENTIALS: WEAPONS

If you choose to carry a handgun on your person, then pick a method that is comfortable and easy to deploy. There are many options so examine them all.

CHAPTER 6
PREPPER FIREARMS AND SUPPORT GEAR

D.I.Y. ZOMBIE SURVIVAL RIFLE BUILD

Zombies are everywhere. All about us it seems these days. I saw one in Wal-Mart just this morning. They can often be seen in the fresh meat section for the obvious reasons of attraction. If you look close enough in your everyday goings about, you will see them, too.

Some consider this whole zombie thing an epidemic. If it is such a serious threat to the security of our world, country, society, our families, and to each of us as individuals, then how shall we defend ourselves against an eventual onslaught of mindless, quasi-human blood-crazed creatures? How indeed?

Think first of all the incidences where the term *zombie* is applied to many things we hear and see nearly every day. It is currently pervasive in marketing angles even at a stretch. Take for example a recent ad by *Grizzly Long Cut Snuff* in a current issue of *Outdoor Life*. The ad line reads, "You think you have enough guns. Then, the zombies come." What in the world that has to do with snuff frankly eludes me (tobacco is not one of my vices), but the point is zombies are given recognition even by marketing gurus (or recent Z-generation college graduates) with the thought it would help sell their products. Scary.

BASIC PREPPING ESSENTIALS: WEAPONS

Since we often rely on the brain source *Wikipedia* to fill in many of the gaps and voids of things we generally don't understand on the surface, then we should naturally expect the Wiki-Org to have a dissertation on the term *zombie*. And indeed they do.

"A zombie (origin Haitian Creole; North Mbundu) is an animated corpse resurrected by mystical means, such as witchcraft. The term is often figuratively applied to describe a hypnotized person bereft of consciousness and self-awareness, yet ambulant and able to respond to surrounding stimuli."

"In modern times (that's now, I presume) the term 'zombie' has been applied to an undead being in horror fiction, often drawing from the depiction of zombies in George Romero's 1968 film *Night of the Living Dead.*" I've seen the movie. If you have not, then I highly recommend a viewing just for the heck of it.

Hollywood of course has taught us much of what most of us think we know about zombies. Newer television versions such as *The Walking Dead* have further portrayed them in a contemporary light. Now we can all recognize a zombie when we see one.

However, us preppers and wannabe survivalists don't really anticipate deadhead flesh eaters to be banging on the front door. I suppose though that some radical mad scientist at the CDC might conjure up a scenario where a dramatic new disease (or an existing one) could sweep the country creating zombie-like characters. See the 1995 movie *Outbreak* starring Dustin Hoffman.

But what about potential epidemics based on such things as a widespread bird flu, brucellosis, bubonic plague, encephalitis, Legionnaire's disease, COPD, tuberculosis, HIV/AIDS, malaria, Ebola, Hanta, Mad Cow, rabies, leprosy, botulism, elephantiasis, lower respiratory infections, and naegleria. Shall I go on? Google some of those if you really want to break out in a cold sweat or not sleep tonight. Would we be able to quarantine our own families and keep the sick wolves away from the doors?

And just when you thought there was but a slight remote possibility of one of the disease threats above coming to a town near you, what happens in America when the entitlement payments stop? When the "empowered" goes to the local convenience store for today's lunch and the EBT card fails to register, what is their reaction going to be? What if China calls in our loans?

What happens when the check won't come anymore? The public dole shows up at a hospital emergency room near you with a head cold, headache, or twisted ankle and they are turned away because the "Care" program bankrupted the country? Will these people simply turn into hungry SHTF opportunistic thugs and thieves?

When the 47% (more like 52+%) no longer has any government system of support, where will they turn? Perhaps these "disenfranchised" masses will turn on us. They may gang up to rove through all our "affluent" middle-class neighborhoods to take whatever we have worked to secure, stock up, and prepare for *us* to survive.

It's something to think about and is a prime subject around our breakout hideout campfire. It's a whole lot more plausible than an invasion by Hollywood zombies.

Though many preppers may elect to opt out of accepting the value of a defensive/offensive weapons strategy, a great number of us fully recognize the importance of keeping ourselves and families safe from any threat. This could be a gang of thugs from across the country, your county, hometown, or a next-door neighbor bent on taking our food, stealing our gas, prep gear, or endangering our lives in the process.

If you are inclined to have a firearm or more for protection, then this is not intended to be another reinvent of your wheel. Others may be just getting into this and know little to nothing about guns and shooting. Selecting and building a SHTF-Zombie long gun may be as foreign to these folks as creating a survival prep food and gear supply was for me years ago.

Again, this is not new science but intended for new preppers or others creating a backup or veterans willing to consider other formats, configurations, or features to stock on hand. Further, please dismiss any arguments or critiques about the choice selected here for any different one that may be more suitable to others. It is what it is and that debate will never end.

I fully recognize that there are multiple viable options for selecting a solid, reliable platform for a workable survival scenario rifle. Without details, discussions, or apologies my personal individual choice is the Stoner Armalite Rifle (AR) platform clone. I accept the good, bad, and ugly of my choice. So far, it works fine for me.

All that was just the marinade, now here's the steak. My own pick is the AR in 5.56/.223 for all the myriad of reasons and justifications for choosing this base rifle. There are other good rifle choices and other good calibers to pick. I chose the AR platform because of its flexibility to customize it to meet my personal needs and interests in a SHTF-Zombie rifle.

I elected to go with the 5.56/.223 because the ammo is very common (if we ever get over the ammunition drought we're experiencing) and easy to find everywhere as a rule. The magazines are also rather universal and easy to locate in supply.

Also there are virtually endless aftermarket accessories which make certain AR clone versions able to accept all kinds of add-on items to make the firearm more versatile.

These accessories have to be chosen carefully for quality and reliability. Certainly there is a propensity to overdo it. Users have to be mindful of all weight added to a firearm that needs to maintain maximum utility.

Here are the items that I recommended to consider adding to a base AR rifle. Such a rifle version should be "optics ready" and come with a rail system installed. That would allow for the attachment of numerous accessories on top, bottom and either side alongside the barrel:

- Collapsible stock. Adjusts to different lengths of pull for multiple shooters. Check out Magpul products.
- BUIS. Back up iron sights (or synthetic) make for standbys if the optical or electronic sights fail. Double-check those batteries.

- Optical/electronic primary sight. I use EOTech, Trijicon, and Leupold. I like GG&G quick release and Game Reaper one-piece scope mounts.
- Sling attachments. I prefer GG&G, Midwest Industries, and Blackhawk. I like the pushbutton release models, but use loop in and clip attachment.
- Slings. Vero Vellini makes excellent slings of several types. The neoprene material stretches, making rifle carry very comfortable. Standard nylon one-inch slings work hard, but can be stiff and tiring to carry.
- Buttstock cheek piece. CAA and others make these. With a sort of rubberized texture these cheek rests increase facial weld to the stock for more steady and accurate shooting.
- Grips. After trying numerous brands and versions I tend to prefer Hogue soft grips over others, especially stock hard plastic AR grips. I like the Zombie green Hogue grips because they glow in the dark. Think about that. Stark, Magpul, and Blackhawk make excellent AR grips, too.
- Rail light. I use the Nite Hunter lights in both white and green light. These come in different lumens with rail mounts included.
- Charging handle latches. After some use many find installing a larger changing handle or latch alone enhances the ability to pull the handle.
- Vertical foregrip. I personally do not like these, but some do. Try one out to see if it works for you.
- Angled foregrip. I have used this Magpul accessory and it does aid in rifle control in sighting and stability.
- Rail covers. These are essential to keep the sharp edges of rails from digging into hands, slings, clothing, etc. I use Tapco, Blackhawk, S&W, and Ruger.
- Enhanced trigger guards. These are helpful if you have big fingers or wear gloves. I use the Magpul polymer versions.

- Other options. Some shooters like bipods. So far, I have not found a use for one. I use Primos Trigger Shooting Sticks. Mag well grips may help some handle their rifles better. Mag couplers hold two magazines together. These are fine, but add weight. Magpul magazine assists fit over the bottom of a magazine as a grab hold for a quick mag release.

Certainly there are many more accessory options out there and more coming all the time. AR accessories are one of the hottest markets going so keep searching the web and supply catalogs. Read appropriate print media that reviews ARs. Buy quality brands not fads.

Finally I recommend you obtain a good AR cleaning kit with ample supplies of consumables like patches, brushes, cleaners and lubricants. A basic AR tool kit is smart. ARs are relatively easy to work on, but don't go over your head until you can. Buy a good rifle case, too. Stock up on extra magazines, and ammo.

Building a SHTF Zombie Rifle can be fun as well as purposeful. Start with the basic rifle learning to shoot it well. Once accustomed to the mechanism then consider adding components to make it *your* rifle. A good zombie rifle should be a good start to protecting life, limb, family, and all else when the balloon goes up or the drones circle.

In this crazy world in which we live that could come crumbling down around our ears at any moment, if we have a little fun with zombies, I doubt it's really going to hurt anything. As a psychologist, I might submit that adding a little levity to all this prepping could help reduce some of the stresses of it all. That would make us better prepared in the long run.

AND MAY THE FORCE BE WITH YOU

Yeah, I know it is a cheesy reference to the movie series, but with a new *Star Wars* movie coming out recently, I thought it appropriate. More importantly I really think Luke Skywalker would love this new optical riflescope and could put it to good use on his exploits.

To be more specific, let the *Nightforce SHV-4-14x56* heavy duty riflescope be with you. With you in the field hunting, popping long range steel, or even so far as protecting private turf from uninvited guests of all kinds. This scope nearly maxes out the full design potential for an optical scope in its class category. Yes, it is that good. And affordable, too.

This relatively new riflescope from Nightforce has opened the door for new dimensions of optical refinery, but with all the desired features that advanced rifle hunters want, but also in a rugged, rough duty service package.

This is *not* a light or light-duty scope so make sure the adaptation for it suits your needs with an appropriate rifle to mount it on. Skip the 18- to 20-inch barreled carbines, vanilla ARs, lever guns, and feather weight single shots and such. This is a real rifleman's rifle scope and needs to be mated accordingly.

Specification-wise the SHV (Shooter-Hunter-Varminter) is 14.8 inches long and weighs in at 28.5 ounces (1.78 pounds) in the illuminated reticle version. The non-illuminated model is slightly lighter in weight, but the lighted reticle is certainly worth the extra scooch of carry weight. It is not a bother at all.

The objective lens diameter is huge at 56mm or 2.56 inches. Can you say "light gathering"? The magnification range adjusts from 4x to 14x. This gives the shooter-hunter a fully functional comprehensive power range for all field-shooting opportunities. There is also a parallax adjustment from 25 yards to infinity. The tube diameter is 30mm. The field of view at 100 yards at 4x is 24.9 feet. At 14x it is 7.3 feet. Amazing!

BASIC PREPPING ESSENTIALS: WEAPONS

The mounting length is 5.75 inches over all with 2.05 inches to the front and 2 inches to the rear of the center control/adjustment module. The controls include fine click adjustment dials for windage and elevation. The left side controls are for the illumination control, battery compartment, and the side parallax adjustment. These adjustment controls are under a secure screw-down cap with easy grip machined grooves on the caps that can be easily handled with gloves on.

To the rear of the scope is the power zoom ring setting and on the eyepiece end is the special European style Fast Focus Ring. With aging eyes and wearing eyeglasses, I really like the Fast Focus Ring and the +/0/- markings to get a clear and resettable focus adjustment point of reference. This may seem a small detail, but trust me it is of great value for precision focus.

The Nightforce SHV's lighted reticle model gives an added feature of high value to shooting during low light hunts or on cloudy overcast days. The watch-type battery is easy to install and change out under the left side "NF" cap. The illumination control is marked for intensity from 0-11 steps. The rheostat can be rotated with clockwise or counterclockwise. Between each illumination step is an "Off" setting so the dial does not have to be turned all the way around back to a zero-off position.

The illuminated reticle is a red "+" in the center of the crosshairs for a clear definition of the internal reticles. If you have never before used a hunting riflescope with illuminated reticles, then you will become quickly acclimated to the clear advantages of having the center of the crosshairs lighted. It is a sweet feature indeed.

Mounting the Nightforce SHV is just as straightforward as mounting any other high-quality optical riflescope. The trick may be finding an appropriate mount and rings for the rifle you use. This can be a challenge trying to find a mounting system for 30mm rings for a standard hunting-type rifle.

The rifle of choice for this project is pretty unique. I was able to acquire a brand new Browning A-Bolt heavy hunting rifle, almost a varminter-type model, with a grey laminated thumbhole stock. The black matte finish against the stock is striking. So is the rifle's chambering in 300 Winchester Short Magnum, my personal favorite deer cartridge.

The issue in mounting the Nightforce scope came in finding a mount and rings to accommodate it. Nightforce's owner's manual specifically states, "Under no circumstances do we *recommend* the use of turn-in style rotary/dovetail type ring and base designs especially those equipped with windage adjustment." Hmmm, that ruins a lot of options. Options that are not available for a Browning A-bolt rifle.

I am not certain what this recommendation is all about but I guess the factory knows. If this was a combat application or perhaps for law enforcement SWAT use, I might understand, but for a standard hunting rifle I don't know the issues. I do not consider the 300 WSM a heavy recoiling rifle and have never had issues with other scopes using this hunting round extensively in the exact same mounting system I used for this project.

Whatever the concerns are, I am taking that risk because after an exhausting search from numerous supply sources and product websites, the only feasible mount and rings I could find for this Browning A-Bolt are made by Leupold. So, I have my Nightforce mounted nicely and quite securely in a set of Dual-Dovetail 30mm rings and two-piece mounts.

Everything went together nicely and with extra high rings I got good scope clearance above the rifle's action and barrel. There was plenty of front to rear mounting space available to acquire good eye relief and there is no contact with the bolt charging action.

At the range, this setup is punch on. I have always had good success with Browning A-Bolt rifles and in particular the 300 WSM. My other A-Bolt in this configuration has taken 14 whitetail bucks and six does in 21 shots. Hey, the sun was in my eyes on the one miss.

Using my standard choice Winchester ammo for my A-Bolts, I have never failed to get excellent hunting groups at 100 yards with this load. The Winchester ammo is their exceptional Ballistic Silver-Tip rifle ammo in 150-grain Nosler Ballistic Tip bullets. Sighting in at a catalog standard three-inches-high at 100 yards, I have easily taken white-tailed deer out to 200 yards without so much as a flinch. I fully expect my new Browning with the thumbhole stock and Nightforce scope to perform likewise. That test is coming soon now that hunting season is here.

The Nightforce SHV scope is a prime piece of hunting gear. It is precision in every way, getting high ratings in the four Fs: Fit, Finish, Features, and Function. You may think this scope for a hunting rifle to be excessively expensive, but what price do you put on a dead-on kill shot on a once in a lifetime trophy white-tailed buck or bull elk? I thought so.

SCOPE MOUNTS AND RINGS

Regardless of the money you just spent on that new scope for your new or old deer rifle, it matters not its quality if it is not mounted properly in suitable rings. Now is not the time to go cheap. You bought a good rifle, a better optical scope, so also buy some high-quality rings and mounts to set the rig up right.

All scope rings are not created equal and neither are receiver mounts. Some name-brand scope makers also manufacture rings and mounts for their scopes. It is a reasonable expectation that, for example, if you have a Leupold scope to mount that their own mounting system is the best available. Of course, those mounts also fit all kinds of other scope-mounting applications.

In fact, for regular bolt-action-type hunting rifles, I only use two brands of mounting systems. One is Leupold and the other is the DNZ Reaper one-piece milled mount and rings. For AR rifles I recommend GG&G quick release or bolt on rail mounts. But those are just my preferences.

Generally I prefer steel to aluminum, being old school, but even I have to admit that there are alloys out there now that can hold up to anything. Cost and reputation count big. A $19 set of rings may not be too high-quality, so buy a good brand. The same is true for the mounts.

Buy rings and mounts of the same brand so there is some degree of match. The rings have to fit precisely into or onto the mounts. Mounts usually come in one-piece or two-piece versions. Both are reliable scope bedding systems, just make certain a one-piece mount does not interfere with cartridge ejection. This should not be an issue with a high-quality system.

Basic rings come in one-inch or 30mm. It is obvious to match the scope tube size to the ring size. Some 30mm rings also include a set of Delrin® inserts to size these rings down to fit a one-inch scope. I tend to stay away from these inserts, because a scope can slip inside of them, but they generally do work fine.

There are also all kinds of specialty rings and mounts for special applications. Some are extension mounts that allow extra forward positioning of the scope. Some rifles like Ruger® have integral rings that fit right on the rifle's receiver. The new Thompson-Center Icon rifles also have a milled Picatinny-type ring mount engineered into the rifle's action. Just make sure whatever mount and rings you buy that they were made for your specific rifle action.

A lot of things can be screwed up when mounting a scope. It isn't rocket science, but it does have to be done right. This is what I do. First get out the right tools and screwdrivers that fit the screws, hex heads, or Torx®-type screws.

Next, I degrease the mounting holes on the rifle with a cotton tip dipped in alcohol or gun solvent. Likewise I wipe down the mounts, rings, and all mounting screws. Let them dry.

I run the mount screws into the rifle mounting holes before putting on the mount. With each screw a dab of gunsmith locking glue is a good idea. Turn the screws down tight for each mount. Alternate between screws when tightening them down. Don't overdo it. The worst thing that can happen is breaking off a screw in the rifle mounting hole.

For the rings, I like to run some 400-grit sandpaper inside the rings to clean off the bluing or stainless finish. Mount the rings according to the manufacturer's instructions. Make certain the rings are mounted or turned square to the receiver so they won't bind on the scope. Then take the top half of the rings off to complete the job.

Once the mounts are set and the ring bottoms are on, set the scope down into the lower half of the rings, put on the top part of the rings, and turn all the screws down loosely. At this time you must adjust the eye relief, or the distance from the eyepiece to your shooting eye when the rifle is mounted to the shoulder. You also have to square up the crosshairs so that they are not tilted. There are tools and fixtures available to assist with these tasks.

When these adjustments are set, begin to tighten down the scope ring screws. Go slowly, alternating between the screws, but do not overtighten. Check to make certain the crosshairs are square to the rifle's action and bore. You should be ready for bore sighting and some shooting at the range to zero it in precisely.

OPEN-SIGHTED HUNTING RIFLES

When I started deer hunting back in the early '70s in Missouri, there was not one deer rifle in camp that had a glass riflescope mounted. Every rifle had open sights, including my Winchester 94 in .30-30. I think it was a good thing we all learned how to hunt using open sights, not optical sights. We acquired some basic gun sighting skills that I fear most hunters today never learned.

The other popular deer rifles of the day were the new Remington 742, the Remington 760 pump, Savage 99, Marlin 336 lever actions, and a host of old military rifles including German Mausers, 1917 Enfields, 1903 Springfields, and British Enfield 303s. All of these rifles sported classic iron open sights. In those WWI and WWII days only snipers had scopes.

We all learned to hunt using open sights. Of course, our orientations to sighting our rifles and then hunting in the woods and field were considerably different than what we see today. We actually learned to sight in our rifles and hit our targets to more limited ranges than what is common today. I sighted in my Winchester 30-30 at 50 yards and don't recall ever shooting at a deer past that range.

We learned to hunt deer more deliberately, taking more time and care in our shots, but with different shooting styles and tactics, too. Since deer drives were common back then, too, some learned to "snap" shoot running game or a deer quietly slipping through the woods to escape the march of oncoming hunters beating the woods aka European styles of game hunting. Our opens sights had to be right on the mark. Open sights are much easier to line up quickly than trying to get a scope on target at a short-range moving target.

Looking down the barrel of a rifle over open sights takes some getting used to, especially if as a youth hunter you first learned to shoot via an optical riflescope. I hate to say targeting is easier with a scope, but then it actually is. The range estimation part and accordingly the sight holdover is still the pivotal element in accurately hitting any target.

Open sights are well, open. The most common style is a ramp front sight, hooded or not, and the rear sight can be a "V" groove dovetailed into the barrel, or an adjustable affair allowing for some windage and elevation changes. The "V" groove type may be fixed or drift adjustable with a hammer and punch. Truly adjustable rear open sights most often use a screw adjustable feature to loosen to move the sight or to "screw" it side to side via the threads in the screw.

Other types of open sights may be found on rifles as well. Some may just have a blade that can be moved up or down for elevation. Sometimes, but rarely it seems, the front sight could have adjustment features.

Of course there are the precision-type "peep" sights and also some of the ladder-type sights found on classic rifles like the Sharps buffalo-type rifles or 1885 Winchesters. My Pedersoli 1874 Buffalo Sharps has a ladder-type-sight that is adjustable via a long Vernier thumbscrew that runs the disc peep up and down for elevation changes. This type of sight is also known as a Creedmoor Sight. It is a very precise sight, as is the rifle.

I took a buffalo with this .45-70 buffalo rifle several years ago on an 1880s-style re-enactment-type hunt in Kansas near the old original "buff" trails from back in the day. We camped in earthen dirt wall dwellings, rode horses, and knee crawled to within shooting ranges of the buff herd. It was a bucket list hunt I will never forget.

Learning to shoot really well with open sights takes some practice and getting used to. First off, don't expect to achieve the kinds of small target groups you might be used to getting with a good riflescope. While an inch-sized group may be the norm with a scoped deer rifle, one with open sights might do well to hit two to three-inch groups and those at 100 yards might be really tough to get.

For this type of open sight targeting, I use the old 10-inch pie plate rule of thumb. If I can print bullet holes in a 10-inch paper plate at whatever range I am seeking, then I consider that adequate to take down a deer if aiming for the vital shoulder for a heart/lung shot. With practice most shooters can achieve this type of rifle accuracy with open sights. It certainly gives one a greater appreciation for hitting targets than when using a 9x magnification scope.

Again however, I am talking much more reasonable shooting ranges using open sights than with a scope. I keep open-sighted big game (white-tailed deer) hunting shots to 100 yards or under. I have no hesitation shooting a deer at 200-300 yards and have killed elk out to 400 yards, but that was with a cartridge capable of handling game at those ranges, and also using a high-quality scope sighted in for such longer ranges. That was with lots of shooting practice at the range, too.

Using open sights on a hunting rifle also implies stalking within closer shooting ranges than per usual when using a scope. I think this is part of the hunting skills that hunters today are not learning or practicing any longer.

It used to be called "still hunting" although the hunting approach was not "still." We would sneak up on game in the woods by moving very little at a time but very quietly over a long period of time. We have simply collectively lost this skill over time by hunting out of elevated tree stands or enclosed shooting houses. Frankly it is easier to shoot a deer off a solid rest of a shooting house window with a scoped rifle, than slipping up on a bedded deer in the woods. It's a lost art of hunting for sure.

So, it might be quite uncommon to run across a deer rifle these days fresh from the factory with only open sights, but there are some still available in this configuration. If you have an interest in a throwback hunting style, then try using a game rifle with open sights. It might well give you a whole new perspective on hunting and a greater appreciation for the sport of it as well.

ROCK RIVER'S HEAVY DUTY (L)AR

I am an unabashed fan of the AR rifle platform. I wasn't always though. I grew up on lever actions and bolt guns as a hunter. I killed my first white-tailed deer with a Winchester Model 94 in .30-30 I bought from J. C. Penney in Columbia, MO for a whopping $66 new in the box back in college in 1971. Wish I still had that rifle.

Sighting in a hunting rifle was a trip back then. Basically it was go out back, set out a cardboard box at some unknown distance and see if you can hit it. I did. That was good enough for all the other hunters in the group.

The next day I killed my first buck from a sitting distance of 10 steps. I have to admit I was sound asleep against a huge oak tree when my subconscious heard a twig snap. When I opened my eyes, there stood a little buck staring me down. I had the trusty Winchester lying across my lap, so I simply raised it up and shot the buck. It was a nine-point, non-typical with every bit of a 10-inch inside spread. I was so proud, I still have the rack mounted in my office today.

Many, many deer hunts have come and gone since those early days and I have managed to graduate up to a number of much more sophisticated deer rifles including several different Remington 700s in various calibers, a couple Brownings, and a selection of Ruger M77s including a recently found rifle chambered for the nearly gone .358 Winchester. It is a real thumper.

I have even taken up deer hunting with handguns, which is another whole adventure I should write about someday. I have to admit that handgun hunting certainly adds a new dimension to the challenges of deer hunting. If you have not tried it, then you definitely should.

But then more recently in the past five years I have been bitten by the MSR (Modern Sporting Rifle) bug. The "MSR" moniker was devised to be "PC" on the part of the firearms industry to camouflage the fact that hunters are actually using AR rifles to hunt. I'm OK with that because much of the stupid press in this country upon seeing an AR in a hunting photo or video assumes the firearm is a machine gun and comments negatively about it. We know better.

My first successful deer hunt with an AR came in Oklahoma on a whitetail hunt at the Chain Ranch sponsored by Smith and Wesson. We used their model MP-15s decked out for hunting in .223 Remington. Though some may think the .223 not enough for deer, we were using special ammunition manufactured by DRT. I proved it was plenty to take down the largest racked buck I had ever harvested with one shot. Others were equally successful on that hunt, taking some big bucks and several huge wild hogs.

So, to say that I was hooked on the MSRs would be an understatement. Since then I have been hunting with a wide variety of ARs. These rifles have been in various configurations from standard fare to highly accessorized versions with copious add-on features including tactical electronic optics or high powered scopes, various sling options, and other user-friendly components.

After all this AR utility, I got the urge to upgrade my power in the field. I think I saw my first Rock River Arms rifles at the annual SHOT Show in Las Vegas. Since then, and after considerable research, I acquired the use of a RRA rifle configuration known as their RRA-LAR-8 Elite Operator. This is an upscale model with many added features that I deemed desirable. I was right.

From the Spec Sheet the RRA Elite Operator is chambered for the .308 Winchester offering both substantially more terminal power than the 5.56/.223 but also heightened terminal ballistics on whitetailed deer or larger game. The .308 is the "bomb" for long-range accuracy, but even in a short-action AR platform this cartridge is a reliable, proven round for hunting.

The RRA-LAR comes complete with the RRA Advanced Half Quad Free Float Handguard with three ladder rail covers. The lower receiver is forged and the upper is forged A4 with forward assist and port door. It has a 16-inch chrome moly HBAR barrel with a 1:10 twist and is cryogenically treated. The gas block is combined with a RRA flip front sight. The rear rail can accommodate any variety of optical or electronic sights, or even an open flip-up rear sight.

The trigger is a two-stage with a Hogue® rubber pistol grip. The buttstock is the RRA Operator CAR stock. Its angled side panels are very comfortable against the cheek. The rifle weighs in at 9.1 pounds and the overall length is 38 inches. It is fairly compact with the collapsible buttstock, but it is not a lightweight firearm. Keep in mind though it is a .308 platform. Add a scope and a full magazine and well, you get the point. It is heavy.

The rifle ships in a RRA plastic case, broken down into its upper and lower units, one magazine and the owner's manual. Go on line immediately and order some additional magazines including a 5- or 10-round hunting mag. I think the new "plastic" magazines are available now.

In practical use terms, I have both hog and deer hunted with the RRA-LAR. The weight of the rifle tames any harsh recoil off the bench, but be sure to wear good ear muffs. I use a Vero Vellini® wide top neoprene sling on this rifle and it carries well with it. I had to move the flat rail cover forward to protect the sling material from the sharp edges of the side rail cutting into the sling and fraying it. The neoprene of this sling softens the carry of such a heavy gun in a first class manner. I highly recommend the Vero Vellini® slings on all hunting rifles.

This is a very accurate rifle using common .308 hunting ammo. I have tried Remington, Winchester, and Hornady ammo and it performs well with all three. It is still open to debate whether to run this action wet or dry. All I can say is if you run it dry, in my case, the action hangs up consistently. A couple squirts of a good tactical lube, cycle the bolt a couple times to coat, and it is off to the races.

You'll get a laugh from this, but do not drop a loaded mag on the ground. On one of my hunts, the mag slipped from my hand and landed in the white sand of the gulf coast area. That stuff sticks like glue. Despite blowing, blowing, and wiping down each round separately before reloading, the sand still got into the action. Yes, it was a mess. It cleans up well with a proper kit.

Carrying the RRA Elite Operator on a first deer hunt, it performed exactly as per my expectations. I was hunting from an enclosed shooting house 10 feet off the ground, which is a favored mode of deer hunting here in the south.

At near dusk a doe walked into the food plot I was overlooking and fed for half an hour. Later a small buck jumped out of the woods line and the doe bolted away with the young stud in tow. Almost immediately a nice eight-point stepped out of the same gap in the woods, walked away a few yards, and began making a scrape and licking an overhead branch. That was his last.

Both the RRA rifle and the Winchester Silvertip ammo did their jobs. The mark was made much easier via the use of a Trijicon® AccuPoint 1-4x24 scope with yellow illumination. I call this a tactical-type hunting scope, but it is perfect for mounting on an AR platform. Once that reticle point gets on the target zone of a deer you are definitely dialed in and ready for venison steak.

Is there anything about the RRA-LAR-Elite Operator I don't like? Not really. It is heavy but I accept that. Again the Vero Vellini® sling comps its weight. It is an extremely sturdy rifle, and geared for hunting or other more serious work as well.

VERTICAL GUN RACKS A HANDY DEVICE

Honestly, there are some outdoor products and super handy gear items out there that simply don't need 750-plus words to describe them. However, some of these "I wish I had thought of that" items I think my readers would really appreciate knowing about.

A couple of gun shows ago, when I was working my buddy's Glock booth, I noticed a new vendor on the show floor that I had not seen before. I walked by the display several times and talked to the booth operator, Mike Nussbaum, owner of Vertical Gun Racks.

Mike showed me his new entrepreneurial idea in the form of uni-body aluminum-formed gun racks in various designs and configurations. The aluminum strips are formed then coated in a heavy rubber coating that is super soft and thus protective for the guns stored on them. The racks are pre-drilled so they can be attached to flat surfaces virtually anywhere.

He has models that can be mounted in horizontal applications as well as a rack that is attached to a wall, etc. vertically that holds a long gun, rifle, or shotgun. This up and down orientation is a space-saving mode that can be installed in gun rooms, closets, workshops, laundry rooms, next to an exit door, or, like I said, virtually anywhere. The horizontal mount models can be put in all sorts of handy, accessible places around a house, shop, garage, barn, or wherever.

The first thing I thought of was putting one in the deer hunting shooting houses at our hunting camp, but obviously my thinking was

BASIC PREPPING ESSENTIALS: WEAPONS

way too narrow. These racks are really designed to make a firearm readily available for self-defense or property protection uses.

Now, Mike will be the first to tell you, and there is a statement on their materials, that they do not suggest or recommend the storing of loaded guns on their racks. I suppose it is like having to wear a helmet on a motorcycle or ATV, but then lots of people don't do it. For safety purposes it is best not to have a loaded gun on these racks.

Without forgetting or slighting their other rack, Vertical Gun Racks also makes a pistol rack. This is a rubber coated "hook" that once installed will hold any handgun, revolver, or semi-auto ready for quick access as well. You simply place the trigger guard over the lower part of the bar and the barrel rests against the hook bar.

Pricing for the gun racks including one vertical rack is $19.95. Pistol racks sell for $14.95 and a Camp Bundle of 10 in a pack goes for $150.00. You can get additional details about Vertical Gun Racks at www.verticalgunracks.com.

As Mike says, "These gun racks go from practical to tactical," which certainly fully describes the options these racks offer. They are made in the USA and intended for gun storage and a space-saving solution. I think if you'll try one of these you will want to order more.

CAT-SCANNED FIREARMS IMAGES

Looking for a really special Christmas present this year for a firearms enthusiast? This is the ticket then. A local physician, radiologist by profession, is not only an active Class III firearms shooter and gun collector, but has developed a unique hobby and sideline.

Dr. Houston Hardin does CAT scanning images of guns. With medical imaging being his professional business and shooting guns his favorite pastime, I guess he decided this was one way to combine the best of both worlds. Turns out his work with producing CAT scan "prints" of various guns is pretty darn neat.

He recently sent me several samples to preview. I got one for the M-1 Garand, an AR-15, Kimber 1911 pistol, a Mossberg shotgun, and a super duper SCAR-17 assault rifle. He has completed the scans of dozens more models of various firearms and continues to add to his catalog listing of prints for sale.

So, how do you think Dr. Hardin got started with this idea? "Several years ago I had a firearm frame returned from a repair. I was out of the office the day the package came, but the delivery man mused that it must be golf clubs. I don't play golf and my staff knew that, so they got the idea to scan the package to see what it was. What they got was the first image of a gun receiver. That sent off a light bulb in my brain. I was on my way," says Dr. Hardin.

"I began experimenting with various CAT scanner parameters to optimize the image quality. Each gun requires its own special settings. Next I worked with a graphics design specialist to clean up the background artifact junk in the scans like the scanner table. None of these prints are photo shopped in the least." They are real CAT scans of real guns.

"Moving forward with the idea I had to create several generations of the scans to produce a really good black and white image to make prints. At first I was just making prints for friends or personal use, but it eventually turned into a full blown project to produce finished prints for sale."

Dr. Hardin continued, "So far I have scanned about 75 various firearms including all original WWI guns through numerous modern arms widely available today such as the Kimber 1911. Each firearm goes through a 3 month process from an initial scan to the production of the final prints."

Houston also told me that others have produced "x-ray" prints of guns, but nothing compares to these CAT scans created on a $750,000 CT scanner. "It's been an interesting mixture of my work with my primary hobby of collecting guns. I only wish I had started on the project sooner as there are vastly more firearm models than I could create prints for in a whole lifetime," says Dr. Hardin. Indeed, what a unique hobby.

On the prints that I was sent, the internal mechanisms of the firearms are really the first thing I noted. It is amazing to see all of the mechanical parts inside of a gun and how they fit together. In a sense it is like seeing the "skeleton" of a firearm.

"As a collector of M1 Garand WWII or Korean War rifles, I was absolutely blown away at the detail shown by these scan prints. It is amazing. I finally have gained a whole new perspective on how all the intricate parts fit together. It has created a whole new realm of understanding of my favorite rifle for me as a collector," says Larry Coleman, a retired police chief, an avid gun collector and shooter.

So, if you collect guns, or have a shooting room, den, or office where such a unique scan could be added as a piece of art, then these firearm CAT scans make a great conversation piece. More details and print ordering information can be found at www.xrayguns.com.

ALTERNATIVES TO THE AR

Believe it or not, there are other semi-auto rifles besides the AR. Without a doubt the AR platform is one of the best all-time best sellers. It has not only been embraced (for good, bad, or ugly) by the U.S. military and law enforcement for decades, but the everyday consumer as well. It is now manufactured by over 120 gun makers worldwide and the list is growing.

It is a fun rifle to shoot at targets, useful for many types of hunting, and is a good solid choice for survivalists and preppers, ranchers, property owners, bug out patrolling, home defense, and a myriad of other applications. But, while the AR rifle makes a good selection for all kinds of shooting work, it is not the only choice out there.

Top on the list of alternatives to the AR rifle and its .223/5.56 chambering is probably the Russian-created AK. Typically we think of this rifle as the AK-47 military base model, but there are numerous other configurations as well, not to mention other origins of manufacture, too. The AK is definitely a hard-working rifle and the 7.62x39 is a proven round as well.

Over the years some accessories makers have designed some ways to replace the stamped metal top cover with ways to supply a mini-Picatinny rail for scope mounting. Some other aftermarket accessories are available, too. AK ammo is readily available. Russian-made steel-cased ammo is fairly cheap in more than one way. Prices for these rifles have escalated over the past few years, mainly due to fluctuating supplies.

Other manufacturers that make various models of "assault"-type semi-auto rifles include Heckler and Koch, FNH (Fabrick-Nationale), Auto-Ordnance-Thompson, Sturm-Ruger (Mini-14), and Beretta, CZ among others. Some of these are carbine-type rifles that use pistol cartridges like the 9mm and .45 ACP, but are still worthy of consideration. I probably missed some, too.

Companies like Springfield Armory (M1A) and DSA (FN-FAL remakes) make upscale rifles that handle the .308 Winchester or the 7.62x51. These are heavy-duty, heavyweight rifles, but control the .308 very well. I saw recently where some new maker is trying to bring out a new copy of the CETME/HK93 in .223. That would be a welcome addition.

Though the AR is a top seller, there are other viable options to consider, so shop around.

AVOID JUNK GUNS WITH A PASSION

Those new to the prepper and survivalist movement interested in obtaining firearms for self-defense or bug out protection are often drawn to the wrong guns. By that I mean the quality, not always the type, brand, or caliber. There are still a number of really poor made firearms in today's marketplace which should be avoided at all costs,

BASIC PREPPING ESSENTIALS: WEAPONS

no pun intended. It is often the cost factor that leads many down the wrong pathway.

I was reminded of this firsthand while working a recent gun show with my dealer friend. Gun shows are the opportunity to buy, sell, and trade, often on both sides of the tables. Many guns come down the aisles from people wanting to sell or trade up. Quite often these guns are of low quality and of very little value though the original buyer may have paid too much for it to start with. Others carrying a family heirloom often are mistaken that their wares are worth hundreds or thousands of dollars. They usually are not.

A seller approached our table with a small chrome-plated handgun, which is usually a red flag to start with. I looked at the gun and declined it. I knew the brand name as cheaply made, unreliable to function, and inaccurate if it fired at all. The dealer looked it over and researched its used value on-line with his laptop. He said the gun was worth about $65 at best, but that he was not interested in any gun of that low quality. The man was incensed, of course. He had paid $225 for a piece of junk. His bad. Which, of course, is probably why he wanted to sell or trade it to start with.

So, how does a first-time or a regular gun buyer tell the junk from quality? Most often it starts with the name brand, but assuredly the price. In this day and age, if you buy any kind of a gun for, say, under $300 you are not getting much. The exception might be somebody you know at work or otherwise in a pinch to raise some money. Still, know the brand and know the gun. Then inspect the condition and look for standard evidence of use or abuse.

If you stick with a well-known handgun brand such as Smith & Wesson, Colt, Ruger, Browning, SIG, Beretta, Kimber, and such you will buy quality. Same for long guns from names like Remington, Ruger, Winchester, Marlin, and many others. There are over 1,000 brand names of ARs now, so pick established, known makers. Beware of ARs that may be homemade by parts assembly in garage workshops. Seek out a good gun value reference book to learn more or perform an Internet search.

Next comes the inspection of the specific gun if it is used. Another good reason to buy NIB, or new-in-the-box guns. Then you just have to research the model, features, caliber, and functioning characteristics such as striker fired, hammer fired, single stack or double, etc. to match your needs, wants, desires, and expectations. You can better learn this at a shooting range or shooting course, which is highly preferred and recommended anyway.

Used guns can be a good value at a good purchase price, or you can get screwed. Obviously if you were shopping for a used car, you likely would not pick one with a missing bumper, or the back fender crushed in, seats ripped, or three different types of wheels or hubcaps. Today you would want a Carfax report to review its history. Get my point?

Look at used guns in this same light, because there is no "Gunfax" and dealers after all are trying to sell what they have regardless. Buyer beware is the watchword.

If the firearm is overly worn, scratched up, stock cracked, sights missing or bent, bluing or other finish highly worn or just looks "rough" then it was probably abused and not well-maintained. Are the weapon's screws intact with screw heads not turned out? Does it look like a shade tree mechanic has been tinkering on it? If so, avoid the potential trouble and move on.

Is the gun clean? Why dealers put a dirty used gun on their shelves or tables is beyond me. If a revolver, check the cylinder for burned powder and a lead build-up. Same for the barrel. Use a bore light, which is simply a special flashlight or clear tube to focus a beam down the barrel. Inspect the barrel for signs of wear. Lands and grooves should be well-defined and sharp, no chinks or chucks, scratches or dark spots. Inspect the muzzle crown for damage.

If a semi-auto pistol, check the chamber and barrel for dirt, grime, and lead. Make sure the magazine is original and locks up. Double-check the safety mechanisms on all guns. Truth is, if a gun looks worn or overly used, then it probably is, but it could still be mechanically sound and safe. Just be careful.

On rifles and shotguns, check the bolt heads, chambers, and loading mechanisms as well as the safety. Cycle the bolt and then bounce the gun on the floor with the safety "ON" especially with shotguns. If the firing pin releases at this test, pass on it. Inspect scope mount holes on rifles to see if the original factory screws are in place. Has there been a scope mounted but it's now missing? Inspect the barrel as suggested above. Look for heavy copper or leading fowling in the barrel which means the owner never cleaned it. Again such a gun could be recovered, but be mindful.

The main point here is if you are just starting out in the buying of a gun(s), then use caution with used guns and, for new ones, just buy a well-known and recognized brand and model. What you don't need is a gun that does not function well or breaks down with little use.

BMS MACHINE CUSTOMIZES ARS

A lot of things are generic these days. That goes from medications to base models of new vehicles to AR rifles that come off the shelf. Now there is nothing usually wrong with these products, it is just a "one size fits all" approach. Haven't you ever craved an AR rifle designed with your own specifications, characteristics, features, and accessories? Well now you can have an AR customized just for you.

BMS, or Bryant's Machine Shop in Jackson, Mississippi, is producing custom-made AR rifles according to the desires of the customer. BMS is a Type 07 FFL dealer and a Class 3 Manufacturer, so they have the creative talents and manufacturing skills to produce just what the shooter wants in a truly custom rifle.

Their AR platform choices allow the customer to choose their particular upper and lower units decked out as they like. They can pick from numerous choices for handguard types, materials, textures, and profile. They offer a wide variety of custom coating colors as well as custom engraving. The buyer can pick parts choices and even request a custom serial number to be applied to the rifle.

BMS's ARs can be chambered in .223, .22-250, 300 AAC, the 6.8 SPC and the .308 Winchester. If a customer is interested in something else, they need only to convey their wishes to BMS owner, rifle designer and manufacturer Ricky Bryant.

Another product that can be added to any AR is a BMS proprietary suppressor. In particular, their .30 caliber suppressors are all stainless steel with "K" baffles in a standard length size of 8-¼ inches. A suppressor can be purchased for around $650 plus the required $200 NFA tax payment with the appropriate application and approval from BATF. BMS can walk the customer through the whole process and assist with every step.

The guys at BMS are big time hog hunters, and they use their own custom rifles with fitted suppressors to shoot the pigs with minimal muzzle noise. This gives them the opportunity to field test their products.

BMS also makes a really cool integrally suppressed Ruger 10-22 rifle in .22 rimfire. These rifles carry a 16.5-inch barrel for light weight and portability. There is also a new take-down version coming out, which ought to be extremely popular.

BMS details can be viewed at www.bryantscnc.com or you can contact them at 601-922-1937. I suspect if you can dream it, they can make it.

DIALING UP YOUR SHTF SHOTGUN

A good shotgun is an absolute "must have" for any survival weaponry cache. Even if you consider building the most minimalist SHTF gun kit, then a shotgun has to be on the list. A plain Jane version can give you basically the same protective coverage, but in today's market of accessories galore, a good smoothbore can be turned into something really great.

BASIC PREPPING ESSENTIALS: WEAPONS

Start out with acquiring a good quality basic shotgun platform. We all have our personal preferences, but with over two million built and in service, the classic Remington 870 pump is hard to beat. They come in a variety of models from austere to fully decked out, including models with shiny, lacquered finishes and highly blued steel.

There are a couple 870 models with laminated wood stocks that would seem even better suited to withstanding the elements longer. There are many arguments for the pump action shotgun, but a good semi-auto can be just as good if kept well-maintained. I would pick the 12-gauge.

Shotguns today can be customized and transformed into a more or less "tactical" version with the addition of some good accessories. Again, take a long, hard look at the aftermarket landscape to see what all is available to bolt on a shotgun to enhance its delivery system and overall utility.

First on that list to consider would be a side-saddle ammo shell holder. These can be as simple as an elastic wrap pulled over the buttstock or even with a sling available with shell loops. Another good option is the bolt-on shell holder that can be assembled along the flat side of the action. All you do is pull the two trigger action screws and insert the new ones through the shell holder. These can be bought to hold five or six extra shells for easy, quick access.

Another good accessory to add is a flashlight bracket up front. This provides illumination under low light conditions. Look into alternative sling attachments, too. On top add a glow sight front and rear. Synthetic stocks are good but not absolutely necessary. Other, perhaps shorter barrels are worth considering, too. One for hunting and one for defense.

The shotgun is an essential survival tool. It provides an authoritative threat with just one cycle of the action. Deck one out to suit your needs.

HEAVY DUTY SHTF BATTLE RIFLES

Sometimes there is a real need for more firepower. As big a fan of the AR-15 in .223/5.56 that survivalists and preppers are, a good argument can be made for having a prep rifle delivering more target impact at long ranges. This means the .308/7.62 in a heavier platform than the M4-type ARs that are so prolific these days.

Likewise, you cannot have the discussion about using the .308 Winchester-class cartridge in a semi-auto "assault"-type rifle (bear with me in using that designation, but I have yet to find a better one. Jeff Cooper, please forgive me) without going to an upsized AR-10-type rifle or another version of a heavy battle rifle.

While the marketplace for AR-15s is way overcrowded with various models, versions, and copious copies, the larger big brothers using the 7.62 have a much smaller market-share of offerings. Basically if you want a heavy semi-auto rifle in this category then you are shopping for the Armalite AR-10, the Rock River LAR-308s, DSA SA-58s (FN-FAL), Heckler and Koch Model 91s, Springfield Armory M1As, or clones of these basic engineered designs.

The good news is that this market continues to expand as the prepper-survivalist buyers demands it. Many preppers that I talk to and advise are looking into adding an upgraded battle rifle for extended superior ballistics just in case. And for good reason, too.

If we compare ballistics for the most basic ammunitions for the .223 vs. the .308, the extra power of the .308 is easily apparent. The standard 55-grain in .223 puts out a muzzle velocity of 3,240 fps and an ME of 1,282 foot pounds. At 300 yards the bullet energy is down to 599 ft. lbs.

By contrast, the .308's 150-grain bullet exits the muzzle at 2,810 fps with a muzzle energy of 2,629 foot pounds. However, at 300 yards this 150-grains of lead and copper jacket produces 1,627 food pounds of smack. This is nearly three times the energy of the .223 at 300 yards. Now do you understand why military snipers use the .308/7.62 round?

BASIC PREPPING ESSENTIALS: WEAPONS

These heavy-duty battle rifles are indeed heavier and more cumbersome to deploy. The muzzle blast is much louder, especially in shorter barrels with recoil to match. However, the terminal target end product is worth it.

HOW STRIKER-FIRED PISTOLS WORK

There is a lot of hype in the gun media news and gun advertising these days about "striker"-fired pistols. As a general shooting consumer, you might be led to think this system is something completely new and revolutionary. Many pistol shooters think they have to run out to get the latest and greatest in handgun designs in order to stay on the cutting edge of the state-of-the-art firearm engineering. In fact, striker-fired pistols have been around for quite a while.

Upon the initial inspection of a semi-auto pistol, if it does not have an external hammer to be cocked back prior to firing, then it basically has to have a striker firing mechanism. In general, all pistol fire control systems are actuated by the pulling of a trigger. When you pull the trigger it either releases a hammer double-action or single-action *or* it releases the striker, which is basically an in-line firing pin to hit the ammo round primer in the chamber to fire the round.

Strikers then are just spring-loaded firing pins that run on an axis in-line with the chamber, which does away with the need for an external hammer to be manually cocked or cocked via double-action by the trigger.

A prime example of a semi-auto pistol design using the striker mechanism is the Glock. Obviously there is no external hammer. The pistol is "cocked" by retracting the slide which pre-sets the striker. One disadvantage to the striker format is in the case of a failure to fire initially, a second pull on the trigger will not re-release the striker a second time. The slide must be retracted again. In training practice this essentially becomes a non-issue.

Upon firing a striker-fired pistol the trigger serves to compete the recycle cocking striker mechanism for follow-up shots. When the pistol chamber is loaded or fired the striker is set to "rest" in a partially cocked position. When the trigger is fully engaged, then the striker is released to fire the round.

Many modern pistols use the striker system including the Glock, Smith and Wesson M&P, Springfield Armory XDS, Kahr Arms, Ruger SR series, H&K, Sig Arms, and Walther pistols.

IDEAL BARREL LENGTH FOR CCW

Is there an ideal barrel length for a handgun to carry concealed? One would suppose so, but then there ought to be some flexibility depending on an individual's personal preferences and what works best for carry and drawing the weapon.

If you were to visit a well-stocked gun store with the intent of surveying the selection of typical concealed carry firearms that are available for purchase now, you would probably find most of the guns with very short barrels. Obviously there is a practical reason for this. I mean have you ever tried to conceal carry a Ruger Super Blackhawk with a 7.5 inch barrel? I can hear you laughing now.

With CCW guns being in such popular vogue now, the manufacturers are cranking them out as fast as possible. Every time there is another SHOT Show or NRA Annual Meeting show the gun makers use that venue to bring out even more new models for the consumer to consider.

The two most popular caliber these days in CCW pistols seems to be the .380 ACP or the 9mm. To a lesser extent but still bought by those shooters able to handle it, the .45 ACP still holds its own in sales. There also continues to be a following for the .38 Special in small revolvers. Even some will buy a .22 rimfire a .32 ACP or maybe a .25 ACP, though this later one is not terribly popular these days for self-defense purposes.

Anyway, the point here is that these are all small(er) handguns, all with short barrels in the general range of two to four inches with two to three being about the norm. Nearly all of the newer pistols coming on the market targeted as CCW guns have short barrels, but even so, they do run the range of the two to four inches generally.

Individual shooters have to decide the balance between the various barrel length offerings, choosing the best handling characteristics, sight planes, sight pictures, concealment features, weight, grips, and types of sights on the handgun. Concerns over recoil could be factored in as well.

Undoubtedly, the shorter the barrel the more compact the weapon; however, barrel is just one consideration. There is also the grip length, grip extensions, magazine fit or extension, firearm thickness and other gun features. This is why shooters have to shop for their CCW with these critical aspects in mind.

KEEP AN OPEN MIND ON OPEN SIGHTS

Fewer rifles these days come from the factory with open sights. There are exceptions though. The Winchester 94s, Marlin 336s, 1894s and 1895s will have open sights. Some rimfire rifles, bolt guns, lever actions and semi-autos still come equipped with open sights. Big bore African double rifles sport open sights.

A few big game hunting rifles by Browning, CZ, Remington ADLs and BDLs, 7400, 7600, Ruger, and Weatherby plus a smattering of others still offer open sights. ARs, AK, HKs and other modern sporting rifles can be equipped with backup iron sights.

One would think then with the seemingly diminished offerings of open sights that shooters have altogether abandoned them in favor of optical scopes or the newfangled electronic red dots or lighted green crosshairs. I cannot honestly say when I last saw a hunting rifle in camp with open sights.

But, before you abandon them altogether, consider the utility of a good set of open sights on many types of rifles. These could be for informal tin can plinking, property protection and security, close quarters shooting, and even for hunting. My first deer rifle. a Winchester lever action Model 94 in .30-30, took many deer before I finally sold it for something more powerful, but still with open sights.

The prime advantage and use for open sights is for instinctive, quick shots on close-in targets. In some sectors of the country the art or science of still or stalk hunting is still preferred. This means creeping slowly through the woods or cover, hoping to jump a deer or pop a quick shot on a coyote or other target. Open sights are ideal for this type of "snap" shooting.

I envision the classic Wisco or Nor-Eastern hunter, dressed in their time-honored red plaid wool coat and a knee-high pair of rubber bottomed boots slipping along the edge of a "popple" Aspen grove wetland slough with that lever-action, open-sighted rifle at port arms. They could snap shoot a deer at 10 to 50 yards with pure instinctive shooting.

If you are a prepper or survivalist, I highly recommend not to get too dependent on glass scopes or battery-power-driven sights. Give yourself the extra option of some good iron sights you can shoot quickly and accurately with a moment's notice.

KNOW YOUR GUN

So, do you really *know* that new gun you just bought? Or even that one you bought last year but still languishes in the cabinet or nightstand? Have you even bought ammo for it? Dare I ask if you have shot the new gun or even practiced with it regularly? If not, then you really don't know your gun and that could be a huge problem down the road if you ever actually have to use it under pressure.

Knowing your gun implies a lot of things. Now, you have been shopping local gun dealers' show cases, gun show tables, the neighbor

BASIC PREPPING ESSENTIALS: WEAPONS

next door has a nice pistol for sale, or you meet some dude off of Crazy List in a dark alley behind the mall to buy a gun. First, know what you are getting and make sure it comes with everything the factory intended to include.

Over the years I have seen and witnessed some weird gun things. I once was going to buy a new Marlin 336 deer rifle, and when it came in the buttstock wood was raw, not even sanded or finished. I have seen magazines missing out of new gun boxes, and many times the owner's paperwork was left out, gone, taken, but just missing. Don't buy that gun.

When you get the gun, after handling it, gripping it, shouldering it or whatever, check everything in the box. When you get home, what is the first thing to do? Before you load it, shoot it, or even work the action, read the owner's manual completely at least twice. You must know right up front exactly how the new firearm is supposed to work. Do not assume anything about this new gun even if you are a self-proclaimed gun expert. It may not work like you think it does.

In this regard learn how to properly load the gun, chamber a round and by all means make sure you understand the safety mechanism(s) and how to unload it. Learn how to clean and lubricate the firearm and disassemble it, then put it back together correctly. When finally at the range, try out several brands of ammo to find out what the gun likes best and shoots most accurately. Every gun varies, even ones of the exact same model.

After you are completely familiar with the new gun, shoot it often. And don't store it away in a fake sheepskin-lined gun rug or case.

CHAPTER 7
PREPPER GEAR

DISTINCT ADVANTAGES OF THERMAL IMAGING

Admittedly, seeing thermal images in the stark of night is cool. Even in a limited product test mode, my mind was going wild conjuring up the multiple uses for such a tool. I was given the opportunity for a month to check out the FLIR Scout PS series handheld thermal imaging unit. I wish it had been the open hunting season.

Without a doubt, thermal imaging brings a whole new dimension to viewing live animals or other objects emitting heat that are picked up by the Scout. The prospects for using thermal imaging are endless when it comes to hunting and other applications.

The more obvious uses for having thermal imaging capability include game management/herd surveys, wildlife observations, security, law enforcement applications, livestock farmers locating lost cows, campers and hikers traversing trails at night, tracking for downed game, and heat loss applications around residential or commercial buildings.

For hunters, the uses should seem more obvious. Walking or riding hunting areas at night just to survey game in the area has its

BASIC PREPPING ESSENTIALS: WEAPONS

own benefits. Anything that emits a heat signal will be picked up by the FLIR Scout. Checking out deer and other game would be awesome.

For hog hunters checking out baited sites at night, thermal imaging is perfect. It also works great for spying fields, food plots, and open trails for active game spotting. The Scout PS will spot man-sized targets out as far as 500 yards. It also works through light fog, smoke, or dust.

The Scout PS is a lightweight, handheld device similar to a monocular optical scope. The unit is held securely by an adjustable hand strap like a compact video camera. It is easy to use, hold steady, and very portable at only 12 ounces. The unit recharges via a supplied USB cable.

The unit itself is ruggedly designed covered in a soft, cushion-like material that will take field bumps, but more importantly is really easy to grasp even when wet. The design was well thought out for the user in terms of ergonomics, and gripability, if that is a word.

The operational control buttons are on top of the unit in a row. Once the user learns which button does what, it becomes second nature to turn on and adjust. The FLIR website has excellent tutorial video guides for learning to use the PS Thermal Camera.

The PS series thermal unit can "see" heat in two modes including *white hot,* and *black hot.* In use I preferred the white hot mode. This simply means that the heat signature showing up on the screen is white as opposed to everything else being white in the background while the heat object picked up is black. It's a matter of personal preference I guess. The white hot mode seemed clearer to me and easier to see detail.

FLIR Scout units are in such high demand for product testing by writers that I only had a short time to use the unit in the field. I took the Scout to my hunting camp in Central Mississippi on an overnight trip. We built a huge fire in the campfire ring out in front of the cabins to give me a base of operation to work from.

Then I set out to walk down the main camp road. I switched on the unit and changed the imaging to the *white hot* mode. The PS immediately began to pick up heat signatures that I would not have expected. Though it was 9 p.m. the tree line around camp was still emitting heat from the sunlight of the day. That was cool to see.

The campfire of course lit up the Scout viewing screen in red hot color. I could also easily see my campmate sitting in his chair by the fire. This was from over a 100 yards away. I picked up all types of heat escapes from the cabin as well.

I continued to stroll further from camp down the road. Next I spotted a small animal on the ground roughly 50 yards out in front of me. The image was so clear it was easy to determine the animal was a rabbit. Its entire body was shown in mostly bright white in the viewer, but more particular the rabbit's eyes were a brilliant white.

I crept up closer and closer until I was within 10 yards of it. My footsteps on the gravel road must have finally startled it, as it ran into the woods. Neat. Next I heard a bird chirping up over my head, and when I put the PS on it, it too appeared glowing white. This proved to me just how sensitive the FLIR-PS is.

Alas, I did not spot any deer with the unit, but if they had been out, I am quite sure their heat visibility would have been spectacular. I learned enough with my limited testing time to appreciate the value of having the capability to read thermal images in the wild at night. Just imagine scouting for deer or hogs with a unit like this.

At a retail of $1999, the FLIR PS would be a welcomed piece of gear for an outfitter, a game manager, or private landowners/individual hunters wanting to assess the status of game in their areas. The Scout PS would also be helpful in spotting trespassers or poachers where they should not be. Also, if game were lost or wounded, the Scout would be great for those tracking efforts after dark. Thus, the FLIR Scout is certainly a recommended piece of high-tech gear every hunter or landowner could use.

BASIC PREPPING ESSENTIALS: WEAPONS

HIDEAWAY GARMENT BAG IN PLAIN SIGHT

The cliché saying is "hiding in plain sight." Preppers need to be sensitive to being able to hide critical gear, supplies, and SHTF stuffs away from prying eyes, thieves, or anybody else that might be snooping around your house or storage areas. This is why we advise keeping quiet and discreet about your prepping efforts, and keeping equipment, gear, and support stuffs out of public sight.

This goes so far as keeping the garage doors down when you are home, or doing work around the house. This includes locking storage rooms, closets, and other storage areas from "friends," family, or guests coming into the home. Using unmarked, locked storage containers is another way to "hide" stuff.

There are any number of ways to secure and hide your SHTF weapons and at least one stash of a short rifle, handguns, ammo, extra magazines, and other related support gear in one hiding place. Here is a review of a new piece of gear designed specifically to conceal your first line of defense gear virtually in plain sight just in case of an emergency. This is a good piece of gear for either a Bug In or a grab and go bag out the back door.

As Skinner Sights owner Andy Larsson says, "Who steals clothing" when thieves or undesirables raid your house? Whether or not you are an official prepper or just interested in hiding some critical guns, magazines, ammo and other gear in a place that is not likely to be discovered either by a break-in or any sort of ravaging during a SHTF episode, you'll be interested in this.

The whole idea behind the Skinner HTF Garment Bag is to provide a unique place to "hang" several guns and other gear out of sight from anyone, virtually hidden in plain sight. Walk into your closet; what do you see? Clothing hanging on hangers from a support rod is an exceedingly common sight in any house or apartment. Shirts, pants, suits in bags, coats, and other items just lined up.

Nothing hanging in a closet should appear out of context or secretive. Think about it. If you were pillaging through a house as a common break-in thief or some ganged-up zombie during a SHTF, are you going to check out the sizes of the shirts and pants hanging up in a closet? Would a dress suit be high on your list of stuff to steal? I don't think so.

So this is the primary premise behind a common-looking garment suit bag hanging in a clothing closet or hallway coat closet among many other suit or clothing storage bags. When the closet door is opened, it should just look like an ordinary closet with nothing out of place. But hidden there could be a secret cache of emergency self-defense gear.

Upon initial examination of the Skinner HTF Garment Bag, once it is unzipped to reveal the internal design features, you simply have to say "Wow." I took a while to really study the inside layout of this new piece of gear. To simply say it is well-designed and completely thought out is an understatement. Upon building it out, inserting all my gear items, and hanging it up in the closet, I am not sure there is a thing I would change.

Everything about this product is heavy-duty. The bag hanger is super heavy-duty despite being made of molded plastic. One would really have to abuse this hanger to break it. It was created to support the weight of all the gear this bag will hold, so it is definitely up to the task.

The bag's zippers and pulls are heavy-duty and designed for frequent, problem-free use. With twin zippers from the top and the bottom of the bag, the user can decide which zipper orientation they want to use that is best for them. I have tried both zippers positioned at the top and then to the left side so they could be unzipped in opposite directions simultaneously. You can play with this to see how you like the closed zipper placement best.

Let's review the inside features from top to bottom. The heavy-duty hanger slips through a double-sewn opening in the top of the bag for hanging the bag up on a closet rod. There is a sort of an overlapping cover as well that is part of the outside zipper closure of the bag. In the very top of the bag is a horizontal pouch for holding a flashlight with a hook and loop flap closure to secure the light. The pouch is not overly large so you may have to try several different flashlight models to see which one fits well.

On the left side of the bag is a position for a rifle such as an AR or AK up to 40 inches in length. The buttstock fits down in the bottom of the bag in an open slip pouch and the handguard/barrel is secured by a Velcro® strap at the top. There is ample room here for a scoped rifle with a magazine inserted if desired. The muzzle of the barrel fits up under the top of the bag cover.

On the right side opposite the rifle containment are two removable holsters held in place by Velcro®. These are positioned one on top of the other with the holstered handgun secured by a strap with a heavy-duty locking clip buckle. The holsters look to be mainly designed to hold pistols, but I inserted a .357 Smith and Wesson N-frame revolver just to see how it fit. The weight of the revolver caused the holster to sag a bit outward, but upon tugging at it, I do not think it would become detached or come lose. Each holster also has a dedicated attached pouch to hold one magazine with a flap closure.

Just to the left of the pistol holsters are pouches to hold six additional pistol magazines with flap closures. Further left of the pistol magazine pouches is a vertical pouch for holding a large knife with scabbard or possibly another piece of gear, perhaps another larger flashlight.

Below the bottom holster are two accessory pouches to be used for a variety of small gear items. These pouches could hold extra lose ammo, revolver speed loaders, or perhaps even a small concealed pistol like a Beretta Pico, KelTec, or Kahr handgun. They could also be used to hold critical medications, extra batteries, a pocketknife, some power bars or whatever.

At the bottom of the bag are three pouches to hold rifle magazines. They are long enough to easily accommodate 30-round AR magazines with no issues. These are also secured with fold-over flaps using Velco® to hold the mags in place.

Sewn down the center of the bag is a nylon strap with the function of giving the bag additional stability and form, I would imagine. Keep in mind when this bag is fully loaded it is going to be pretty heavy, especially with all the magazines loaded to capacity.

Once the bag is geared up, the outside of the garment bag is zipped closed to hang up. I think a small padlock could be put on the zipper pulls. Also quite unique is the fact that the bag can then be laid out (on the bed or table) and folded in half to be secured by two sewn-on external straps with snap closure buckles. The outside of the bag also has two sewn-on wraparound carry handles to tote the bag like a piece of luggage.

Unobtrusively hanging in the closet? Yes, but note upon unusually close examination of stuff hanging in a closet this bag will stand out somewhat. Clearly the heavy-duty hanger could be noticed as something different. The external carry straps and wrap straps attached to the bag could be noticed. I recommend when hanging the Skinner Garment Bag in the closet to fit it in between other suit or coat bags to further obscure its presence. I seriously doubt it would ever be noticed hanging among all the other clothes or bags.

This bag is constructed of heavy-duty cordura black fabric with tough stitching. I have seen photos of the bag in dark green. I think the black is a better choice as it tends to blend in more appropriately as a garment bag. The retail pricing is $179 and can be ordered direct at www.skinnersights.com.

So, whether to hide in a closet or fold over for a grab bag, the Skinner HTF Garment Bag is an alternative way to secure an essential cache of weapons, ammo and support gear out of sight in plain view.

BUYING A BASIC AR RIFLE

There is a big misconception about buying an AR rifle. I hear it time and time again when a newbie prepper or budding survival

BASIC PREPPING ESSENTIALS: WEAPONS

enthusiast queries me about building a weapons cache for self-defense, property protection, or bug out security.

These folks have been watching too many "shooting" television shows, or reading too many articles in the new genre of tactical shooting magazines. While some of these publications offer good information and gun reviews, they tend to focus on high-end ARs and all the accoutrements to be hung on their rails. I read these magazines and enjoy the reviews up until I see the $3,500 price tag. That's ridiculous.

For a beginner or a bare bones shooter there is no need to start out with a new AR rifle that is all decked out. In fact, if you are new to shooting or if this would be your first AR-15, then do yourself a favor and avoid a rifle platform with all the installed bells and whistles. That is, until you are fully versed in the use, shooting, and maintenance of the AR rifle in a Plain Jane version.

What constitutes a basic AR? Go shopping to a well-stocked gun shop with ample inventory of AR-15s or attend a large gun show just to look and ask questions. A basic AR rifle can be of two types actually: a copy of the more or less original AR format with a carry handle, "A"-frame front sight, triangle fore-end and a fixed buttstock. There are many variations.

The other likely version is the M-4 type with a collapsible 4-6 position stock, flat top, optics ready, but may come with open sights and a two- or four-part rail fore-end or a standard fore-end with "A"-frame front sight. Which to buy is based on preferences and prices. Expect to pay around $600-800 for the most basic, new in the box rifle.

I do vote for the M-4 versions with three Picatinny rail sections on top and sides for future expansion options. Go with the open sights first, adding optics later. For now, forget the flashlight, laser beams, hand stops, vertical grips, and other accessories.

Go basic. Shoot the gun, disassemble it, clean it, carry it, and get confident shooting it, changing magazines, and enjoy it. You can always sex it up later.

CHOOSING A SHTF AR RIFLE

Remind me again how many flavors of ice cream there are at Baskin-Robbins? Looking at all those tubs of ice cream reminds me of checking out the well-stocked rifle racks at a big gun store for AR rifles. New preppers or survivalists getting into the gun game must be overwhelmed by all the choices, not to mention the toppings.

Budget considerations come first. I called my bud at Brandon Arms just to get an update on the price range of AR rifles at his shop. His go from a low of $769 to right at $1500 on the top end. If you do any shopping for ARs, attend gun shows or read magazines like *Firepower*, you would be shocked that some ARs are priced up over $3000.

Personally, I think that is ridiculous and absolutely unnecessary for the average prepper or enthused survivalist. You can do much better with some dedicated and judicious shopping around. And a high-quality, little-used, pre-owned AR should not be out of the realm either.

Now I can hear you saying right off the bat that you can build an AR from your shoebox of parts for under $500. Good for you. But I am talking about the *average* AR rifle buyer for prepping, hunting, or just fun time shooting. These folks are not gunsmiths, don't have the tools or skills, and probably never will.

At last research there are over 100 manufacturers of AR rifles. Every time I pick up a shooting magazine, I see a new AR rifle by a maker I never heard of. This can be problematic, too, so beware. Buy an AR rifle or any other product for that matter from a known entity with an established reputation for quality, delivery, and continued product service.

A quick *Google* of any brand name can get you reviews by real users to at least provide a benchmark of comparison between rifles. I use this method all the time to see what others think of a certain rifle, optic, ammo, accessory, or whatever. This goes for household stuff

BASIC PREPPING ESSENTIALS: WEAPONS

as well as any prepping gear or supplies. None of us have money to just throw away on junk.

What brands of ARs are good and worth checking into further? Sure, I have my own biases, but there are some AR rifles that I trust are of good enough quality and reliability to warrant their consideration for an investment. Any omission of a brand on my part does not automatically mean I would not buy the rifle. Frankly, as mentioned, there are too many to keep up with.

For a heart and soul AR, I can recommend LMT rifles, Colt, Smith and Wesson, Ruger, Stag Arms, Rock River, Alexander Arms, Anderson Manufacturing, Daniel Defense, Les Baer, and Yankee Hill. Certainly there are other companies making good ARs, I just have not inspected or used them personally. If any maker of an AR would like us to field test their rifles, then they can offer one up for us to use. We certainly cannot afford to buy them all.

So, what do you look for, shop for, and what features do you need in an AR rifle? I think you can still get a really bare bones AR rifle, new, for $600-700. I would plan on at least spending $1,000-1,500 for a decent rifle, optics, and some accessories you might wish to add, even later on.

A basic AR would be a black rifle in .223/5.56 with a 16.5-18-inch barrel with flash hider. It would be a flat top, optics ready model with an installed Picatinny rail on top or maybe a full rail system on all four sides of the barrel. Ideally it would have good BUIS (back up iron sights) on it. The rifle should be aluminum or alloy, not polymer even in a lower unit (personal bias there).

It can have a fixed stock, but in my user opinion, a six-position M-4 type stock has much more utility. There should be several optional sling attachment points or fixtures from standard metal sling loops to plug in holes for push-button sling attachments. A forward assist bolt plunger and shell casing deflector are nice as well as an ejection port dust cover.

Prime accessories to consider later might be extra *Magpul* magazines. I use exclusively *Hogue* soft feel grips. Maybe add a quick detach or folding forward vertical battle grip or a forward hand stop. You'll want rail covers to protect your hand from the sharp rail edges. You might want a bipod later. Maybe add a flashlight. Get a good discreet gun case for low profile and rifle protection.

Scopes, optics, electronic holographic sights, red dots and such will be covered later. Ditto on mounting systems for these. Know that these can easily cost more than the rifle investment.

So, pick out a good, basic AR and learn to shoot it and maintain it. Add all the neat accessories as you need and can afford them. We saved up a long time to get our fender skirts back in the '60s. We were just tickled to death to have wheels back in those days.

EXTERNAL BELT GEAR RIGS

When the soldiers left the ships to fight in that big war to end all wars, the troops were all carrying a webbed belt around the outside of their coats or jackets. This webbed belt carried a wide variety of accessory pouches for ammo, weapons magazines, medical supplies, a canteen, maybe a holster for a 1911 Colt .45 and other optional gear items.

The external webbed belt kept the gear weight well-distributed around the waist and easy to access. Without carrying these immediate-need items on the pants belt itself, the soldiers would not have their trousers weighted down or pulling excessively at the waist. Also this web belt could be quickly detached to set aside; however, these rigs were usually carried at all times.

Today, preppers and survivalists would do well to copy this gear carry mode themselves. In fact, such rigs are once again finding favor with outdoors enthusiasts from hunters, campers, hikers, and survivalists working around bug out camps. These external belt rigs can be customized to easily carry needed items that are used often or that can be reached or deployed quickly.

With a little planning and thought, such an outside carry belt can be easily designed and outfitted. What gear should be added to such a rig? Make a list then narrow down the choices.

Start with a heavy-duty belt. Some still like and carry the old military surplus webbed belts and these can work with the proper accessory attachments. Better yet is a thick leather belt that will not bend or bind with a load. I bought a double-layered leather 1.5-inch wide belt recently off the rack at Cabela's. It is super stiff, but will become more pliable with use. It has a good brass buckle.

Make sure whatever belt you get is large enough with enough adjustment holes to fit over outer clothing, including light jackets as well as heavy coats. It may be best to wear a coat into the supplier or retailer to get a proper fit over the outer garment. Try on different styles to see what seems to work best.

So, what to hang on such a belt? The first thing that comes to mind is a sidearm weapon in a holster. This of course can be any handgun that you use confidently and have practiced with often. Likely you wear this outdoors, so if working on a farm, ranch, bug out camp, or similar environment, you may want a handgun with substantial enough power to dispatch varmints or other intruders that might invade your space.

The most common choices that most will pick include a 9mm or a .45 ACP. Revolver shooters will pick a .357 Magnum (for which .38 Special ammo can be used), a .44 Magnum (with .44 Special ammo) or perhaps a .45 Long Colt. Obviously other choices are available, too.

You handgun choice can be fitted to any number of holster types and styles that suit your uses best. Pick a heavy-duty, durable holster with good gun retention. A safety strap is not a bad idea, because when working outdoors and such you do not want any likelihood of the firearm dropping out of the holster or being snatched out by a tree limb or vine.

Next besides a weapon would probably be a good camp knife. The blade choice should be something between a hunting knife, general purpose Bowie, or heavy blade that can do some chopping along with regular field cutting tasks. An ESEE #6 comes to mind. If you want or need a pocket-knife-sized utility blade or two, then carry one of those, too in a smaller scabbard.

Now comes all the options that preppers, farmers, or other outdoors workers might choose specifically for the kinds of field work they are performing. It might be a hatchet or small hand ax, a mini-first aid kit with meds, a canteen, compass, cell phone w/case, ammo pouch or pistol two-magazine pouch, bear spray, other accessory pouches (forestry tape, bright eyes, paracord, insect repellant, small digital camera, snacks and nabs, fire lighter, flashlight, multitool or other gear items). The balance in picking these items is not to unduly weigh down your belt rig.

Where to wear or use this belt rig? Obviously, outdoors, but such a rig could be worn while working inside and out around the bug out camp, farmhouse, barn, or other situation. It should be an easy take-along when riding an ATV, UTV, or even a horse or tractor. The rig would be ideal for walking the property to inspect fences, gates, and for security observation. The belt rig would be good for hiking trips, too, assuming such carry is permitted on the trail.

WWII soldiers found great utility in the everyday carry of their gear around their coats with a webbed belt. It spread the weight around the waist, but gave immediate access to needed items. Preppers and survivalists can adopt this type of rig for many uses and to perform a variety of tasks. Be creative in how you design your belt rig so it becomes a real go-to gear carry option.

GOTTA LOVE THE LANGUAGE ARTS OF TACTICOOL

Tacti-what? This new generation of weapons (guns) enthusiasts has apparently been raised on a steady diet of acronyms, mil-spec designations, Emojis, hyphenated non-words, street hype, gun shop slang, ballistic mumbo-jumbo, combat colloquialisms, and

BASIC PREPPING ESSENTIALS: WEAPONS

non-descript barrack's rattle they now call *tacticool*. Where exactly that term came from is anybody's guess. And so, why bother?

This past year I directed a federal grant from the U.S. Department of Labor. The learning curve in terms of understanding the millennial language used in the grant as well as in conference calls from the administration "sorority girls" as Michael Savage calls them is amazing. You can't even understand most of the time what the hell they are talking about. They assume everybody is versed in the gov'ment lingo, but few of us are. It's an epidemic.

A lot of this new language evolution has been rolling over into survivalist, prepper, and weapons material for the past few years. It has been coming on pretty fast, and it is hard to keep up with and to fully understand. For this reason, I write this tome to warn preppers to take time to fully digest this language and to not be fooled or misled by it. Much of it is just window dressing trash talk and marketing hype anyway to lure in a segment of readership that has been slipping away via iPhones and other technologies.

So, take text descriptions like "I 'lased' the target" or "check the scope's reticle algorithms" with a grain of salt. The last time I "lased" anything I was taking a nap with the lap pooch in my recliner. That explanation almost anybody can understand.

Some newbies to prepping have recently sent emails asking what this term or that term means, and though I chuckle to myself, I try to fully explain what I know if there is in fact anything to be known. Most of the time I can muddle through an author's depiction of some new product review or technique trying to decipher the millennial lingo.

The really fun part is thinking from my higher education background that most of these 18-22-year-old Generation Z (for zombies I guess) junior warriors probably can't even read the articles to begin with, much less understand them. The next generation up that served proudly in the sand dunes or other scrappy places probably do "get it," since I think this is where most of the terminology coinage comes from. So be it, just read it with interest and bypass the blow.

GUN LOCKS

Why is this so hard to understand? Gun in the house? Kids in the house? Kids have friends over that have never been around a firearm? Yet, curiosity has definitely killed or severely injured more than the cat.

If you own a gun or guns, then secure them. If you do not have a gun safe or a lockable cabinet of any kind, then any gun that is in a place that can be found by anybody should have a gun lock on it. You have to plan ahead, but use common sense to ensure a balance between having a personal security weapon ready and handy as needed, but secured away from those that have no business handling it.

Time and time again, you read in the papers or hear on the television stories about young children or even teens that ought to know better playing with a gun that was loaded but not secured. These "accidents" often result in fatalities or serious wounds or lifelong debilitating results.

An employee in my old office has a son now paralyzed from the neck down due to playing with a gun at home with a friend that went off. The bullet lodged in his neck, and his life and that of his entire family was ruined forever. Thousands of questions and "if's" followed the incident.

One could strongly argue parental negligence in such cases. First, any gun in the house should either be unloaded, or locked up, or secured in a fashion that only the primary user can gain access to it. A gun should not simply be "found" in the home not secured. This is the responsibility of the parent or guardian.

The irony is that securing a firearm is so easy to do. In fact, many manufacturers of pistols and long guns now provide a simple gun cable lock in the factory carton as shipped. Instructions are also provided on how to open the action of the firearm, slip the cable through the open bolt or action, then insert the cable end into the padlock connected to the other end of the cable. Done. Hide the key or put it on a keyring and your gun is secured.

Hornady also makes several pistol safes that function in a variety of ways to secure a handgun at home, work, or in a vehicle. Today, there simply is no excuse for not locking up a gun.

JUDGING THE INHERENT VALUE OF A HANDGUN

What process do you go through in the selection of buying a new handgun? Of course, this applies to the purchase of any new gun for your use, new in the box or used. Undoubtedly there are many strategies used to pick out a new gun. For some, including me, and perhaps unfortunately, it is often an impulse buy. That is the worst way to buy a new firearm.

The smart way to approach buying any new gun for survival, prepping, or hunting is to ask the simple question, "What am I going to use the gun for?" Even if I am buying a shotgun, will it be intended for self-defense inside the house, or for quail hunting or duck hunting? While one gun could work for all three if you change out the barrels and/or the choke tubes, it might be difficult to find a suitable universal firearm for all purposes. What are your thoughts on that?

If you think you want to shop for an appropriate concealed weapon, then what do you buy? Are you planning to take a handgun safety shooting course, a concealed weapons course and the enhanced carry class to obtain a legal carry permit? In that case, you are not going to want a Ruger Redhawk or likely a tiny Beretta Tomcat either. So, first figure out the task and then shop for the appropriate tool to accomplish the goal.

Once you know for what reason you are buying a gun, then narrow the choices by the appropriate design and features. If you want a semi-auto pistol, are you going for a hammerless striker-fired gun, or one with an external hammer? What caliber will you choose, and this is a highly critical choice. Do lots of research and get lots of advice on this one.

There are well, too many choices of guns to buy so picking one is tough. Visit large, well-stocked gun shops, or gun shows, to see the widest variety in brands, designs, styles, and features. Handle as many as you can since a pistol must fit your hand well in order to shoot it well. If you can, locate a shop or shooting range that rents guns to try out several models. The value of a gun is if you shoot it well, and it does the job.

CHAPTER 8
PREPPER STRATEGIES

HANDGUNS FOR DEER

Handgun hunting for deer is everything it was intended to be. Is it the ultimate challenge for a firearms deer hunter? Yes. Is it the extreme test of hunter stalking and shooting skills? You bet it is. If you are looking for a new deer hunting experience to reinvigorate your hunting, then give handgun hunting a go.

It was a cold morning just north of the Desoto National Forest near Brooklyn, MS, which is a grand piece of public hunting ground. Spitting rain, cold rain, and the deer were not moving. I was walking back to the truck and as I rounded a corner along the edge of the woods a doe was standing just out of the woods line about 75 yards away. I dropped to my knee.

I steadied the open sights, took a couple more deep breaths to calm down, and pulled the trigger. The doe fell, but I did not see it. I thought I might as well have been shooting a black powder rifle. That nanosecond when hot burned powder gases exiting the muzzle turned into a cloud of fast frozen mist obscured the results of the shot. It cleared and one look through the binoculars confirmed the kill.

That was my first confirmed deer taken with a handgun. I used one of the first Ruger Redhawks in .44 Magnum on the market and not the Super version that came later with the capability to add a scope in factory dovetail mounts. I swapped for that model later, because I was totally hooked on handgun hunting for deer.

If you are really interested in taking up deer hunting with a handgun, you have a number of important decisions to make. First on that list is what type of handgun you like the most and can shoot best. Some may consider the caliber choice to come first. There is a fine line here, so picking either first is OK.

The assumption being if you like the gun, its action, and features, you will practice with it more. That is the key to successful handgun hunting. Unless you are willing to shoot your primary handgun for hunting enough to be highly proficient with it, then this may not prove to be your cup of tea.

The three basic types of handgun actions to consider for hunting are the single-action (Western-type revolver), double-action (Dirty-Harry-type revolver), and the single shot bolt or break-open handgun. I am biased, but for handgun hunting, forget the semi-auto type action. The exception might be a Desert Eagle in .44 Magnum.

Each type in a given firearm of high quality and known reputation will be fine for deer hunting if the hunter will learn to shoot it accurately at a variety of deer hunting ranges, usually 100 yards or less. Handgun hunting is generally a short-range affair.

Be sure to handle numerous models of various handgun types that are suitable for hunting. Try them for hand fit, a comfortable firm grip, handling weight distribution, useful open sights, or capability to easily adapt a scope mount. All the while also keep in mind if blued steel or stainless steel is more to your liking. There are benefits and distracters to both.

Some factory handgun models are specifically designed for hunting and thus will be built to accommodate an LER (long eye relief) type handgun scope. Ruger makes several of these models, and the Thompson/Center single shot can be easily scope-mounted as well.

BASIC PREPPING ESSENTIALS: WEAPONS

In the old days the minimum standard rule of thumb for picking a cartridge suitable for deer hunting was the all-time classic .30-30 Winchester. Given that, realize that only a very few cartridges designed for handguns can achieve the ballistic power of the .30-30 with a general muzzle velocity and energy level of 2,450 fps, and 1,995 foot pounds. These would be the .454 Casull, .460 and .500 Smith and Wesson. As a disclaimer here, I am not talking about shooting a .308 in a single-shot handgun either.

Much has been written on picking an appropriate handgun cartridge for deer hunting. Just do a *Google* on the subject to find more material than you probably care to read. Generally speaking, the primary handgun cartridge choices for deer hunting include the .357 Magnum, .41 Magnum, the .44 Magnum, .45 Long Colt (marginal), and the .454 Casull. The .460 and .500 are equally as suitable if one considers the recoil of these, but they are highly effective deer cartridges for those able to shoot them well.

Handgun hunting can be done on any public WMA or National Forest during regular open hunting seasons. Check the regulations. I think you will find the challenge just what you were seeking.

THREE-GUN SET FOR SURVIVAL PROPERTY SECURITY

Here's the scenario:

"After a harrowing night of driving back roads, dodging roving road blocks, and trying to negotiate through a couple of small towns crowded with wandering looters, you finally wake up in your bug out hideout. The reality shocks you awake. Hundreds of questions pop into your mind. The wife and kids add to the list. Are you safe and for how long?

"After years of prepper planning, you were finally able to buy a parcel of land in a sparsely populated rural area well away from the city where you worked and raised a family. Now after social upheaval and confrontations between local police, then National Guard with bussed-in paid agitators, authorities have lost control. General mayhem is widespread.

"You scan the AM-FM radio for news updates, but learn little. The family is settling in, but are nervous. It is time to venture out to patrol the parameters of your property to get any sense of prying eyes or trespassers. What guns do you take?

"You clocked plenty of hours of research, preparation, purchasing, and training for this day. You have assembled what you feel confident is the best three-gun set of defensive firearms to protect yourself, family, supplies, and property integrity."

The three-gun set was simple. One each: handgun, rifle and a shotgun. With limited resources, his decision was based partly on economy, also on priorities to have each type of firearm available, but also to have sufficient ammunition to support his choices. So, what did he pick?

For his first SHTF pistol, he chose a Beretta 92 in 9mm. It is a well-made, strong, accurate, and relatively easy to shoot handgun. He bought three extra high-capacity magazines, but wishes he had bought more. With a good nylon holster and belt, this is his prime sidearm for now.

His rifle is an AR-15 clone decked out in a defensive style with a red-dot scope and multiple accessories to make it comfortable to use and carry. He acquired 10 *Magpul* magazines, all fully loaded.

His shotgun is a 12-gauge Remington 870 pump with multiple choke tubes and a compromise 26-inch barrel. He bought a case of buckshot and one of No. 6 shot for hunting.

Is this the perfect three-gun set for survival security? It is as good as any given the limitations. The smart part is having one of each type of firearm for multiple applications. What would *you* choose?

ACTIVE SHOOTER RESPONSES

Being prepared for an active shooter incident is even more of a reality after the situation in California recently. What would you do if an active shooter entered the building where you work, or a shopping mall or hospital or other facility? You need to know how to respond immediately to protect yourself, family, friends, or colleagues.

First on the agenda, if you happen to be unlucky enough to find yourself in the middle of an active shooter incident. Act fast. Secure your immediate area. Lock and barricade the doors of the room where you are. If you happen to be in a hallway or open area, then get to a confined room where you can lock yourself in.

Turn off the lights, close the blinds, block windows, turn off radios and office equipment, and remain hidden and quiet. Stay out of the view from any door window if you cannot cover it. Keep all the occupants calm, quiet, and out of sight. Block entry doors with desks, chairs, or whatever is available. Stay behind protective walls and filing cabinets to bullet-proof yourself. Then silence all phones. You do not want anyone roaming the halls outside to hear noise from your room or to otherwise discover your hiding spot.

You can call 911 when the event initiates. When you report an event, tell the authorities your specific location, building name, and office or room number. If you have spotted any of the assailants, tell authorities what you saw. Locations, number of suspects, race/gender, clothing descriptions, physical features, types of weapons, and anything else you can think of.

They will want to know how many are with you and if anybody is hurt. If that is the case, assess the injuries and help as needed.

Once you lock down, stay there as long as it takes for authorities to come to secure the area and do a room by room search and check. Don't panic. If they knock, then be sure to identify them before you blindly open the door. Follow their directions to escape.

Active shooter incidents are usually short-term events and assailants are not likely to spend much time trying to breech a locked and barricaded room. Remember, hide yourselves and stay secure until you are guarded out of the facility.

ALTERNATIVE PERSONAL DEFENSE WEAPONS (PDWS)

In the usual scheme of things, survivalists, preppers, and the general population interested in pursuing self-defense options are most often centering their focus on firearms. In lieu of the immediate availability of a suitable shooting personal defense weapon, then we have to consider other options. It could be that some are just philosophically or psychologically opposed to owning or shooting a gun. In that case, they definitely need alternatives.

There are a myriad of tools to consider wielding for self-defense as needed. Just walk around your garage or storage room to see what you have hanging on the wall. You might be surprised at what you own already that could be used to thwart a zombie or a thug that wakes you in the middle of the night.

As an illustration, I did just that in my own garage. Hanging there on the pegboard walls, or sitting on the floor, are these items in no order of preference. On the floor sits a sledgehammer and a pickax type of tool. There is a four-foot-long piece of steel car axle. A hardwood flag pole stands in one corner, as does a fiberglass extension tree-limb cutter.

On the wall is a standard claw hammer, a ballpeen hammer, a heavy chain cutter, shovels, hoes, and two different kinds of cutting blade machetes. Other tools include long, heavy screwdrivers, both Phillips and flathead types. There is a short-handled garden trowel with triple spikes on one side and a sharp flat spade on the other. That would hurt on a swing to the head.

There are also assorted hedge clippers and trimmers, including one electric hedge trimmer that would cause a nasty abrasion to the

BASIC PREPPING ESSENTIALS: WEAPONS

neck or arm without the power on. A WWII fold-out shovel in its original case hangs there. With the spade folded at 90-degrees, just imagine its impact. There is more, but the point is we can find ample tools around the house to defend ourselves. And this does not even count the kitchen stuff.

Recall one of the most recent episodes of *The Walking Dead* where Eugene faces off with the metal-head guy in the factory? Eugene tries to reach for a piece of steel rebar lying on a nearby table. Excellent choice, though out of reach, so his partner completes the task. The bottom line is we just have to look around and use what is available as needed.

AMMO SUPPLIES

I received an inquiry recently from an outdoor writer contact in Wisconsin; he asked me about local ammo supplies and inventories so I did some checking. At the past two or three gun shows I did take note that while ammo stocks seemed to be widespread and comprehensive, their prices remained sky high with all the usual excuses.

I thought once the "wars" began slowing down that demand for ammunition was supposed to decline on the part of the military, the prime consumer of ammo. But that reasoning seems to continue to fail to hold water. There also seems to be a reduction in the number of reports on various government agencies no longer buying a kagillion rounds of ammunition. Perhaps those reports have just been squelched.

I also perused the shelves of several big box stores a few months ago and found their stocks lacking. I don't know if ammo ordering is the responsibility of the gun department manager or the general manager of the store, but I never could understand how an outlet like a Bass Pro Shops could be totally out of various calibers, brands, and types of ammo, but this was certainly the case in the local store here in Central Mississippi.

Now, after a recent survey of ammo stocks at some big box stores I can report good news on ammo inventories. I checked the local Bass Pro Shops, Academy Outdoors, and a Dicks Sporting Goods store. Two of the three had very good stocks with shelves more full than I have seen them in months. Academy has always maintained the best ammo stocks even during the ammo buying rushes. Our local Dick's has never had much of an ammo stock, but it is respectable. Some Walmart stores have good ammo stocks, while others are dicey.

As I said, prices remain high. Some good ammo buys can be found at outlets like the *Sportsman's Guide* on-line, especially if you wait to buy when their free or minimum purchase shipping deals are on. You just have to shop around to get the best prices, but ammo is just not inexpensive these days. I suspect the manufacturers want it that way.

My recommendation is to start shopping now in advance of hunting seasons to get the selections you need or want. Some pre-hunting sales may save you some bucks so be prepared to buy.

IS THERE A BEST AR RIFLE HOLD TECHNIQUE?

For those of us old school shooters, adapting the new rifle holding and shooting positions takes some getting used to. I was taught how to shoot hunting rifles from an early age; most of us were taught to grip the rifle's stock grip or space behind the action, and put the offhand under the front of the stock fore end section for support. Though this still works, it is the best positioning for an AR?

I started to take note of the new styles of holding and shooting ARs when I surfed the shooting shows on television watching the three-gun shoots and other ARs being shot at the range. Then I picked up on it while reading some of the shooting/combat magazines off the rack. Now, I am trying some of them out firsthand to see if my accuracy improves. What AR holds are working for you?

BASIC PREPPING ESSENTIALS: WEAPONS

Though there are some AR shooters still using more or less traditional rifle holds, the new vogue style is to place the offhand well forward on the fore end virtually as far forward as the shooter can reach. The idea is to lock that arm "into" the rifle to add stability and a faster recovery from felt recoil.

I have seen shooters cup their offhand under the rifle and some over the top of the rifle's rail if it does not interfere with the open BUIS sights or an optical sight option. Take careful note of that. It does seem practical that moving the offhand forward does add forward support to the barrel that should aid in accuracy and moving the rifle to targets or from one to another in a more fluid motion. The hold appears akin to holding a bow at full draw.

Other AR holding options include gripping the forward part of the magazine well. There are even accessory attachments to place there with fingerhold gripping surfaces. Likewise there are devices now to attach on the under rail called *rail handstops* that can be pushed against by the web of the offhand or by the thumb.

There are also vertical grips that can be attached under the rail, too. There are multiple designs of these intended to be held tightly by the offhand to add controllability or stability to the rifle.

The bottom line is to find an AR hold that works for you whatever the style may be. What is your style?

ARE YOUR DEFENSIVE WEAPONS TASKWORTHY?

As a prepper, how comfortable are you that your chosen defensive weapons are the best choices for you? Does your primary go-to pistol really fit your hand all that well? Does the length of pull on your MSR (modern sporting rifle) really fit your torso and arm length for comfortable, read accurate, shooting? Can you pump that shotgun without taking it down from your shoulder for a follow-up shot? If not, then your guns may not be task-ready.

Furthermore, can you manage the controls on your pistol with your off hand? Meaning, if you are right-handed and that arm is put of out commission, can you operate and shoot your pistol with the other hand? Can you reach the safety and mag release on your AR without taking the rifle off target? Can you pull the charging handle left-handed? Easily?

An ideal defensive weapon has to fit and work for the shooter in the most efficient manner possible to maximize effectiveness. This can be difficult or impossible with some gun designs, so accordingly the operator has to train more extensively to get this weapon into action and keep it on-line.

When you begin the process of choosing weapons for self-defense then consider these points in an attempt to find a firearm or three that serves your needs the most.

Sights on any firearm should line up naturally as you bring the weapon to bear. The sights need to be highly visible. If you use electronic optics on a rifle, then choose an illuminated dot or reticle for better target definition. Handgun sights should be white dots or bars, and should be night sights that glow in low light.

As much as possible buy firearms with ambidextrous controls. Safeties on both sides, enlarged slide releases, and magazine releases. Make certain guns using change-out magazines can be quickly released and inserted with ease. Confirm your gun's function will all types and capacities of magazines.

Rifle and handgun stocks and grips should be easy to hold and fit well. Gripping surfaces should be easy to grasp and hold onto. This might mean replacing factory grips with something like a *Hogue* rubberized grip. ARs with adjustable M4-type-stocks can be positioned to fit the shooter's length of pull. Defensive weapons need to fit well to perform best.

BAD TRAINING WORSE THAN NO TRAINING

What with the heightened interest in concealed carry in this country, the permit requirements for self-defense training has generated the creation of a plethora of shooting schools and training courses. Many of these schools are owned, operated, and taught by highly qualified instructors with proven curriculums. However, some could only be classified as "fly by night" with instructors having little more credentials than big egos.

Before you sign up for any self-defense course, or a qualification class to secure a concealed weapons permit, check your state's laws, rules, and procedures for obtaining the permit. Some states may certify and recommend specific training schools or programs to meet their criterion for acceptable training. If not, ask around at the permit agency, ask law enforcement officials, local gun shops, and even search the old Yellow Pages. Do your due diligence to find a qualified school.

What credentials or qualifications should you look for? First ask or check to determine if the course instructors are officially certified NRA instructors. Ask to see the certificates. This means those trainers have attended and passed these rigorous courses to know what needs to be taught and how to teach it.

Certainly quiz any instructor or school official about their training resume and reference list. If they don't have either, then pass it up. How long have they been in the firearms training business? How many classes or participants have they trained? What are their course offerings? Do they work in an established location and facility or are they working out of a backpack in a field somewhere?

When I took my course, I knew the instructor personally. He used a training room at the local Bass Pro Shops store. The firearm's range and course of fire was conducted at his deer hunting club some 20 miles away. It was not the best situation and to the unknowing, it might have appeared a rather sketchy setup. Check these things out in advance so you are comfortable with the whole situation. Again, if in doubt call the permit agency to ask if this instructor or course has qualified other permit seekers.

Ask the course instructor about the classroom course. Is it an approved NRA course with a provided course book or just a bunch of handouts? Again, make sure your comfort level is high on this. Be sure you know the difference between a standard course and an enhanced course and what the shooting requirement are. Bad training is a bad deal.

BOLT RIFLE OR SEMI-AUTO?

There are many choices when it comes to selecting a long gun for multiple uses. Many of the questions and inquiries I get from preppers and survivalists are about their attempts to buy a firearm(s) that can yield effective results for many applications including home defense, ranch, farm or homestead protection as well as hunting for food and predator control. That is a pretty tall order for sure.

After much thought, counseling, and work in the gun-related industry these past 40-plus years, the basic conclusion I have come to is that the rifle preference really boils down to personal choice. I mean in terms of overall quality, reliability, functionality, and accuracy there is not a significant difference between major makes of long guns now, whether a bolt action or a semi-auto.

While the caliber choice may be the first priority, that is no longer a huge issue either because the most popular choices in the .223/5.56 range up to say the ever popular .308/7.62 are readily available in either platforms with numerous brand and feature choices to select.

It would be easy to recommend if all you could afford was one choice, then for sure, I would say the .308 would get the nod. It is fully capable with available factory ammo choices to perform work in protection and certainly for hunting and dispatching vermin regardless of the foot count.

BASIC PREPPING ESSENTIALS: WEAPONS

Though the fight breaks out when you mention the bolt rifle is inherently more accurate than the semi-auto, there are plenty of examples to defy that argument these days. I have been hunting with bolt guns since I was 10 years old, but I have also taken big game with semi-auto rifles including an AR in .300 Blackout and a Rock River LAR in .308, both one-shot kills.

Having owned and used both types of rifles, I do find the simple bolt action rifle easier to use, much more simple to clean and maintain, and easy enough in most cases to mount a scope. I love the quick and easy way to remove the bolt so the barrel can be easily cleaned from the breech end and not the muzzle. Shooting a bolt action is fairly straight forward and easy to train others by controlling the ammo use at the range. Safety mechanisms are usually simple to operate.

Even with bolt guns using detachable magazines, I have had no issues with them loading or unloading or feeding reliably. I use both a Browning A-Bolt and a fine Remington 700-DM without issue. I did buy factory magazines for these to have an extra in the shirt or coat pocket and they have always worked just fine.

Learning to use most semi-auto rifles takes somewhat more training and range initiation. Loading, cycling the action, and engaging the safety can be easy, but takes some practice to perform. Using magazines takes little effort, but sometimes they can be finicky.

I would be first to admit I am not crazy about taking down an AR-type platform rifle, removing the bolt carrier group from the action in order to be able to clean the barrel. I am not fond of all the nooks and crannies that come with most semi-auto rifles either. I will be honest though that this is most likely for me a function of frequency than it is performance of the tasks.

I can envision the GI soldier taking down, cleaning, and re-assembling an AR-15 in the dark just by feel. It isn't that it is so difficult, it is just that I don't like doing it.

Of course some other semi-auto rifles like the Browning BAR or the Remington 742 series and others may be more intricate or detailed in terms of disassembly for cleaning and maintenance. For this, just follow the general guidelines in the owner's manual and don't take them apart any more than is recommended as necessary for regular maintenance.

In terms of scope mounting, most bolt guns simply need the application of an appropriate one- or two-piece mount to be installed on top of the action. Then separate rings are attached to the mount. AR-types mostly use the Picatinny rail to mount scope rings directly to the rail slots or via a one-piece unit incorporating the mount and rings in one-piece ala Nikon, or GG&G.

There are many variations to these themes, but neither is a big issue. Both bolt guns and semi-autos can be easily and securely fitted with an appropriate scope. Some adjustment in ring height might be in order depending on the size of the scope's front objective rings to clear the barrel. This is more of an issue for bolt action rifles but can be with some ARs as well.

It is a toss-up when it comes to rifle reliability. Just by the sheer action from moving in a relatively violent fashion, the semi-auto sometimes is blamed for more breakages than are typical bolt action rifles. Many times stoppages are due more to powder and grime fouling in the gas system than actually a part breaking. Semi-auto actions do get pretty dirty, pretty fast, and therefore often demand more regular cleaning and maintenance.

Having said that, it would probably surprise most rifle shooters to learn that most guns these days are so well-made that breakdowns are far and few between. That is all else being equal. I mean most of us are not on the battlefield firing thousands of rounds in a short time. Don't be fearful of buying either a modern bolt gun or a semi-auto when it comes to reliability issues.

So, when you are ready to buy a long gun (rifle), then shop around. Seek advice, but be careful of the big box gun counter person that worked in Underwear yesterday. Read, study, and compare. Visit a local range and ask questions. Again, it is a personal choice whether you want to cycle a bolt action or pull back a charging handle. Whichever you end up with, be sure to practice with it.

BUG IN SECURITY

You are sitting in your favorite recliner after dinner at the end of a day of work that stressed you out mentally and wore you out physically. All you need is a quickie 30-minute nap to revive you for some work you need to finish in the yard. Before you doze off, you click on the nightly news only be informed that the SHTF has finally hit.

A riot has broken out in a nearby shopping center after a mob swamped a big box store for some electronics sale. Two people were shot and several were injured from a crowd stampede. Now the rioters have moved on to burning cars in the parking lot and are moving down the street breaking into other businesses, looting, and setting them on fire.

The police have been thwarted by the thugs that are more heavily armed and grossly outnumber the numbers of law enforcement on hand. An overhead helicopter camera report shows the advancing crowd of bandits is only blocks away from your neighborhood. What do you do now?

Had you devised and been working on a Bug In prepping plan, you could execute that immediately. Let's assume you did plan ahead. First get your family together and move on the plan. You totally lock down the house, including lowering the metal window shutters and double-bolting all the steel-reinforced entryways. You plug the tubs and start filling them with water. You move extra supplies of water and food plus other SHTF gear storage containers from the garage into the hallway.

You bolt-lock the garage door and disconnect the electric opener. You lower the lights to a bare minimum. You turn on the emergency radio but keep the television on low volume to monitor ongoing news.

You open the gun safe and pull the necessary weapons, handguns, rifles, and shotguns. You grab ammo pouches, ammo cans, and security vests. You place some appropriate guns and ammo at strategic places throughout the house along with flashlights and candles.

Now all you can do is wait it out. Maybe with your house dimmed down, quiet, and looking vacated, any thugs will pass by. If not, you stand ready to defend your position, hope the power stays on, and pray the event does not last long-term. What will you do?

CLOSE-QUARTERS-COMBAT

Are any of us ready for a CQC event? Perhaps many of us fancy ourselves as pistoleers like the boys of *Young Guns*. Most of us will likely die thinking so. I hyphenated the title of this piece on purpose. The point being to separate the components into distinct concepts.

Research on first responder shooting events indicates that most armed encounters are experienced really up close. Up close has been defined as from three to seven feet, often closer. What is your concept of a close encounter? Close your eyes and imagine somebody threatening you from only three feet away. You can almost reach out and touch them. That's scary.

If you are aptly carrying a concealed weapon, can you conceive of having to draw your CCW firearm, control it, extend your arm to point it, then precisely aim it and fire at an adversary that close to you? We're nearly talking powder burns here.

So as you train to carry and deploy your concealed carry weapon, consider shooting for scenarios where the "range" is only an arm's length away at best. That takes a high measure of mental and physical control to train for such events.

Next define quarters. Quarters could be inside your home in the living room, at an external doorway, down a hallway, or other confinement. Is it dark? Do you have a flashlight and can you operate a pistol with it?

BASIC PREPPING ESSENTIALS: WEAPONS

CQ could be in a grocery store parking lot where you are approached by a knife-wielding thug. Maybe it is inside the store when you are in line checking out as a robber jumps over the counter to push the clerk out of the way and brandishes a gun at you.

Such events rarely happen under ideal conditions or situations. You never know what position or corner you might find yourself in. Train for all kinds of situations both inside structures as well as outside, in the relative open or other confined spaces. Learn to work the quarters to your advantage as cover.

What does combat mean to you? In CCW situations combat is not usually a military mode of extended exchanges of fire. You might find yourself with the chance to fire just one round face to face. It has to be quick, concise, purposeful, and accurate. Your life, and others, may very well depend on it.

DIRECT ACTION TACTICS

This term is nothing new. You can Wiki it and get a full explanation and historical overview of what it implies. As survivalists and preppers though, maybe it is time to learn how to engage in such actions to protect our own interests, security, and holdings. Sometimes we just have to step up to the plate to defend our positions and rights.

I have a nosy guy in my residential neighborhood. He is retired so everybody's business has unofficially become his business. He rides around the streets in his van with his big attack dog (which is really a big baby when his wife walks him) looking at everyone's property and trying to catch anybody breaking the neighborhood covenants. Really, it is none of his business.

Once I had my daughter's vehicle sitting in *my* driveway off the street with a "For Sale" sign in the window. This guy stops, gets out of his van, and walks across my property to tell me I can't have a car for sale sitting in my driveway. You can only imagine my response, which I cannot print here. I very directly and quite impolitely told him to get his keister off my property and to stay the blank-blank out of my business. He still tries to stare me down. Maybe he does not understand what the middle finger stands for.

The point is, a time will come during a SHTF, blackout, storm aftermath, or other event, when we will absolutely have to stand our ground. It might be as simple as telling off a bothersome neighbor or brushing a shirt tail aside to display a concealed firearm with intent. It is going to happen.

Now none of us are hopefully going to go out of our way to start trouble. But, then sometimes you have to nip it in the bud before it starts. This is a mindset that preppers are going to have to adopt whether they like it or not. It is like getting bad service at the drive-thru window, at a fancy restaurant, a car garage, or any business where we are spending our hard-earned money. We expect to be treated fairly, honestly, and get what we paid for.

Otherwise we become the abused victim. During a SHTF event, an attitude like that could become life-threatening. We have to learn to stand up for our rights, then defend them.

NEW POTUS EXO DOES LITTLE FOR GUN CONTROL

Depending on your point of view, the recent executive order to initiate new "gun controls" is really just a re-emphasis of laws and rules governing gun sales since the initiation of the Gun Control Act of 1968. However, like so many other American laws on the books, many of the current ones are simply not enforced.

It *is* still federal law that legal gun sales processed through federally licensed gun dealers or FFLs require a background check on the purchaser. There are no sales of guns directly to consumers via the Internet. Those businesses tell buyers plainly up front that all guns have to be shipped to an FFL for legal processing and ATF background checks.

The chink in the armor comes when sales are made between private individuals at gun shows, on the street corner, or in the church parking lot. Laws still restrict some of these sales, for example, to minors, or sales of handguns across state lines.

I work gun shows for a licensed dealer and I see what goes on. A guy walks around the show carrying a deer rifle he no longer uses or wants. Another individual approaches this person to inquire if the gun is for sale. IF there is an agreement on the price, then the gun changes hands without any background checks or paperwork processed. This is a legal sale.

True too, are sales called a "straw purchase" and *they* are illegal. A question on the required BATF Form 4473 asks "if you are the legal buyer of this gun." If that person lies on that form but still passes the required background check, then the sale goes through. It cannot be the judgment of the licensed dealer where that gun may end up later on.

Furthermore, as the AG has stated, the real problem is unlicensed "dealers" setting up at gun shows selling multiple guns without a license for which no background checks or paperwork is completed. Legal dealers want this stopped, too, but again, this has to be monitored.

So, in the final analysis, this new executive order is not going to prevent a legal gun buyer from going off the deep end to commit a crime or murder. These are tough issues with no simple answers. Trust me though, legal, licensed gun dealers are following the law and conducting the required legal background checks whether at their retail shops or a gun show.

FITTING HANDGUNS FOR SENIOR WOMEN

Some time back I worked the grand opening of the Boondocks Firearms Training Academy's Pro Shop in Raymond, Mississippi. I came away with one primary lesson learned. Older women want to learn to shoot. They want to have the empowerment of self-protection. Some of them are just downright scared and want to be able to defend themselves. They want to buy a handgun and learn how to handle it safely and shoot it well.

Accordingly, never let your husband buy a handgun for you. Let a professional firearms instructor appropriately "fit" a proper handgun to your hand's feel and grip. The other thing I learned is that obviously no two people's hands are alike and not every handgun model fits everybody's hands.

Older ladies especially with petite hands or even a slight touch of arthritis have a very difficult time manipulating handguns. Over the course of two days, I probably counseled with twenty ladies that fell into this category. First, they were somewhat apprehensive and scared to handle any firearm, but I found most of them could not operate some kinds of guns.

Some of the ladies could not cock a revolver. Others could not pull a double-action revolver trigger. The strength was simply not there. One lady I worked with could not even open the cylinder on a .22 magnum rimfire revolver to simulate loading it. This inability to do these normal handgun operational tasks was very frustrating for them. Dedicated training and individual instructor assistance would help immensely with these situations.

Nearly universally these senior ladies could not cycle the slide on any striker-fired, non-hammer-operated pistols. After I showed them some of the semi-auto pistols with a hammer and they were able to cock them, sometimes with the palm of the hand, surprisingly most could then pull the slide back. A decocking feature would be best for them.

Some were so frustrated they simply walked away. I was able to coax some of them back for more discussion. In a couple cases I handed them off to one of the range instructors. Hopefully some will come back for a formal class.

Ladies definitely want to learn to shoot a gun. Instruction is the key to their success. They need a handgun custom chosen and fitted. Then they will feel confident to successfully shoot it.

PROXIMITY TO GUN STORES A GOOD THING

A news article hidden away on the bottom of the page in the *USA Today* news revealed an interesting piece of information. It stated that 98 percent of the U.S. population lives within 10 miles of a gun store. This statement was compiled by some national mayors' organization. I feel certain it was presented to be negative. That was the exact opposite of how I took the news.

When I was a kid growing up in a small rural farming community in Southeast Missouri, my access to guns came via the local Western Auto store. For me, just to purview the small inventory of guns was a delight. I am guessing the entire stock was less than 20 guns, mostly shotguns for bird or duck hunting. We had no deer in that area back then.

My dad bought me my first gun at that store, a Ruger Bearcat single-action Western-style revolver. He paid $40 for it. In those days (1955) a box of .22 rimfire ammo was about 50 cents. Besides reading *Boy's Life*, that store was my link and education to guns, shooting, and hunting.

Today, I have access to multiple stores to shop for firearms. Apparently, you do too. I think for those of us shooters, collectors, hunters, preppers, or other gun enthusiasts this is a good thing. Many of these "gun shops" are national big box stores like Wal-Mart or Bass Pro Shops, but lucky for me, there are still several really good "mom and pop" gun shops left in business to cruise up and down the aisles to shop for shooting stuff.

It will never happen in my lifetime or yours or that of your grandchildren, but I guess people will never view guns as mere tools. To me it is no different than going to a hardware store to buy a hammer or ax. Those tools can do great damage, too, in the hands of the wrong person. And such actions we can never stop.

Years ago on a trip to Russia, I met with the owner of the only "gun store" in Moscow. He sold mostly fishing supplies, since he could only stock less than 10 guns at a time. I am thankful we still live in a free country where we have access to firearms within 10 miles.

GUN UP

Today's world is a scary environment. For those of us born before, say, 1960, particularly those raised in small farm, rural communities, life was safe. We never locked our doors at home. We left vehicles unlocked as just a matter of course.

Nobody messed with our stuff. If I rode my bike to school a mile away from the house, when I came out at the end of the day the bike was still there. Same with stuff in a school locker in high school. We never locked them. Of course, back then we did not have backpacks, iPods or iPads, laptop computers, pocket calculators, or any of that common stuff today that gets stolen every few seconds now.

If we hunted, we just crossed the fence in the backyard with a pellet rifle or shotgun and went hunting. Permission was openly granted without having to ask. Try that today. For a while into high school we could keep a gun in the back seat or on a rack in the back window of the pickup truck. Try that today.

In those days, we all shot guns, had guns, or family access to them for the most part. It was just the way we were brought up. Not so sure about youth in big cities or large communities. We just assumed *everybody* grew up with guns. Not so as I found out later when I went to college.

Now much later in life it has occurred to me that even though most prepper types are big on guns, self-defense, and all that it takes to get ready for a SHTF scenario, not everybody is "into" guns or shooting. Many preppers are scared of guns and don't know how to get started with learning to handle them safely and shoot them effectively.

Most preppers that pursue the survivalist course of study realize the apparent need to defend themselves and everything else that involves the use of firearms. They now begin to realize the need to "gun up." It is just part of the reality of the world we face today.

To get started, visit a full-service gun store with knowledgeable personnel. Seek advice on a training course, shooting range, or personal protection course. Choose firearms carefully, then pursue hands-on training to build your confidence and skill set. Practice often and continue to study.

GUNS LOADED IN THE HOUSE

With any type of firearm in the house, gun safety is paramount. This matters during the regular course of daily activities, the stresses of SHTF events, or even during bug out survival circumstances. Plans have to be made in advance for safely keeping loaded guns in the house.

Years ago, a good friend of mine's wife was hosting a birthday party for a friend's kid and his buddies. Alan was a trained pistol shooter, so he kept his Colt Combat Commander upstairs in a bedroom night stand…loaded all the time.

Just by accident, he went upstairs to get something and found two of the kids in their bedroom playing with the pistol. A round was in the chamber, but the safety was ON. Nevertheless, he never got over that incident and no longer kept a loaded gun in his house.

I face a similar situation with a young special needs child still at home. She has never picked up a gun (unloaded) without my supervision and does go to hunting camp with me, but has never learned to shoot. I am very careful about where I keep loaded firearms in the house. I am sure others face similar considerations.

In this day and age of increasing social unrest, home invasions, thefts, and assaults, more and more people are buying guns and keeping them loaded at home. This requires some judicious planning and thought about where to keep those guns.

Another friend of mine who is a self-professed prepper has bought several of the hollowed-out books where a pistol can be hidden inside the cover. He has these placed in several spots around his house for quick access when needed. This is certainly one approach.

Hornady makes a handgun safe called the RAPiD safe that can be set out in the open and is operated by passing a security card over the panel. These pistol safes certainly can work well, but they will be located in just one spot unless you buy multiples of them. However, this does solve the issue of keeping a loaded gun in the house.

Other security locks, cables, and such are available and are worthy of consideration. Long guns have to be dealt with differently so just be safety conscious when keeping all guns loaded in the residence, or a vehicle for that matter.

HOW TO SHOOT OUTSIDE FROM INSIDE

Some things in life are easier said than done. Try changing a flat tire on an incline for example. Get way up on a ladder in an otherwise dim bedroom to install a ceiling fan. Open the hood on your new model car and figure out where the spark plugs are. Some things are better left to a knowledgeable professional. Some things cannot be avoided.

As a prepper, if you are faced with a SHTF situation in which thug gangs are roving the community streets in search of easy prey, then you have to be prepared to defend your position against these worst case scenarios. Likely this will include prying eyes or crowbars seeking to enter your residence for food, water, valuables, guns, and your girls.

This is a scary situation at best, but as a survivalist caught inside, you have to be prepared to throw back any threatening advance. You can only low-pro your existence inside your dwelling for so long. Even tiny candlelight can be seen through cracks in blinds. Smart thugs will spot trash piling up on the street or other signs of occupancy. Get ready. Be ready.

Your last resort is going to be confronting threats from the outside, so you have to train how to shoot from the inside. This naturally causes exposure issues, but better shooting from behind a door or the corner of a window than out on the front porch or behind the air conditioning unit.

Advanced preppers will prepare exits and window coverings to lock down the house from approaching storms, hurricane winds, or other threats. Shooting portholes can be designed into these security covers ahead of time. They might be "T" openings, or crosses like those used in old English castles. You need to see out and be able to aim your firearms through these portals.

If this is not an option then consider shooting from a slightly opened door and how to block/lock a forced entry. Ditto on windows. Work to get as wide a field of view as possible. Train and learn to shoot on your knees. Have pads and ear protection available at each shooting position. Take turns to avoid fatigue.

To learn these skills you can build some training doors and windows like a shooting range. Teach your family to shoot or at least learn to change out magazines for you. Have multiple guns for multiple positions. Shooting outside from inside takes practice. Better to learn now than later.

CHAPTER 9
GUN AND GEAR MAINTENANCE

BEST GUN-CLEANING BRUSHES

If anybody out there really enjoys cleaning their guns, then you need therapy. Well, not really. Some hunters and shooters not only enjoy cleaning their guns to restore accuracy and durable protection, but just like to know their guns are inspected regularly and kept well-maintained.

Top on the list of good cleaning supplies are bore brushes that can really get the job done effectively and efficiently. I have my own thoughts on that after 50-plus years of cleaning bores from .22 rimfires up to the big bore .45-70 buffalo round I used in Kansas a few years back to take an American Bison.

I, too, like the idea of keeping my guns cleaned especially after a shooting session or a long hunting season. Sometimes if my hunting guns were exposed to bad outdoor elements then I will clean them again mid-season or as needed. I always wipe them down at the end of a day outside with a lightly oiled rag just to keep them spot-free from skin contact, dust, and exposure to the elements.

When I settle down at the work bench out in the garage or the work table I use in my indoors man cave, I break out a selection of gun-cleaning supplies to get the job done right. One of the first tools I get out are the bore brushes I will use to clean the barrel, chamber, and inside the action.

Call me old school or whatever, but I only use bronze or brass brushes in the bores of my guns. I have used stainless steel and plastic bristle brushes before, but after years of experience, these simply do not get the job done as well. Why?

Neither stainless steel nor plastic brushes conform as well to the rifling of the barrel bore in order to get into the crevices of the lands and grooves to satisfy my bore light inspections. When I am done scrubbing a barrel bore, I want to see a mirror shine with no grime or goo stuck in the rifling grooves. This can take some stubborn scrubbing and elbow grease.

I also tend to use a caliber-size large bore brush if it is not too tight and, of course, always clean a barrel from the action end, not the muzzle. Use brass or bronze for better results.

BUILD A SHTF GUN-CLEANING KIT

Preppers do not always have to buy everything already packaged in a retail consumer kit. Sometimes it can not only be cheaper to buy just the items you really need but it can be twice the fun to assemble a bug out gun cleaning kit of your own.

Years ago as part of prepping our practice trips to my bug out location, I decided to fashion an easy-to-pack gun-cleaning kit of my own. Those hard plastic or wood gun-cleaning kits are OK, but I wanted something different customized to my own needs and easy to pack.

When I bug out, the two types of firearms I pack out mostly are an AR rifle and 9mm or .45 ACP pistols. So, I gather just the cleaning supplies I need to keep those weapons in tip-top condition out in field situations. Sometimes I add a 12-gauge shotgun. Create your kit accordingly.

Start with a heavy-duty gallon zip-lock freezer bag, the kind with the slide locking feature, not the press-together grooves in the top of the bag. First get small plastic bottles of gun cleaning solvent and a good gun oil or lubricant. Put in a good quality, soft, absorbent gun cleaning cloth, the type that does not leave lint behind every time you wipe gun metal with it.

Then buy one of the coated cable or cloth snake pull-through bore cleaners. Obtain a correct cleaning patch slotted pull-tip, and proper-sized brass cleaning brushes. Get a couple cotton bore mops if you like to use those. Pick out a selection of cotton bore cleaning patches to match your respective gun bore sizes (and save the used ones for your fire starting kit).

Add in a couple of the military-style plastic bristle gun cleaning brushes with both the larger brush on one end and the smaller one on the opposite end. A brass scrub brush is good, too, as is an old toothbrush. If there is extra space in the kit bag, then a really nice tool to have along is a flashlight-style bore light with a plastic optical bore illuminator or a lighted flex tip.

When you bug out either to hunt or escape during a SHTF scenario, you are going to want to keep your guns clean, so assemble your own kit.

CLEAN OPTICS NEXT TO GODLINESS

OK, OK, forgive the reference, or not, but if you can't see through that smudged-up riflescope or binocular, then why have them? Maintaining clean optics is one of the easiest things to do now with all kinds of glass cleaning products for hunters on the market, so why put up with dirty glass?

Glass optics in the forms of riflescopes, binoculars, spotting scopes, and even rangefinders are designed to enhance the human eyesight experience and provide an elevated level of visual clarity to the target being viewed. If the glass is nasty dirty, then you are only defeating the design function of that device. How smart is that? This includes cameras and trail cams, too.

The week before I go deer hunting, I line up all the optics that will be in the field with me for a complete inspection and cleaning. I happen to use a variety of products to accomplish this. First I have a zip-lock bag that I call my "optics cleaning kit." This goes to deer camp with me just like all essential gear.

Inside the kit is a little book of camera lens cleaning tissue paper, a small spray bottle of eyeglass cleaner, some clean cotton patches, a couple of computer-screen cleaning cloths, and one of the lens cleaning pens by Leupold. Used properly, these items will clean and shine any optical glass to a spotless gleam. Keep the kit handy for when you get back to camp after a hunt to remove dust and smudges again.

Attached to a loop on my hunting backpack is one of those handy optics cleaning cloths that stuff up into a holding sleeve. These are cool and I often use them in the field while in my hunting stand if there is a misting rain, or other conditions to dirty my optics. With aging eyes such as mine, I want my scopes and binoculars to be crystal clear all the time.

As a sidebar, I have seen some of the anti-fog products on the market, but have yet to find one that works. One was a little tub of a saddle-soap-looking goo. Don't use that stuff. It is like putty glue and smears around forever. I had to use rubbing alcohol to get it off.

Basically, it is easy as pie to keep your hunting optics clean, so do it.

FROGLUBE EXTREME

Who says that old dogs cannot be taught new tricks? I was the kid that oiled his bicycle chain so often that it often dripped spots on the ground. Of course, with that much oil on the chain and gear, it picked up a ton of dirt, sand, and debris which ultimately caused other problems. The moral of the story on lubrication is to only use just a small amount for the job.

Now I am reading the guidelines and suggestions for using a new lubrication available for gun owners and shooters. This guide says "Remove any petroleum, lubricant, fouling or residue from your weapon using FrogLube Solvent prior to applying FrogLube. DO NOT mix lubricants in the same weapon." Well, hush my mouth. Apparently I have messed up again by not paying particular attention to what lubes I am using, or mixing use on the same weapon. Let that be a lesson to you.

FrogLube is a part bio-based, non-petroleum, non-toxic, non-flammable, environmentally friendly system designed for military and law enforcement use, fully complying with the U.S. Army directive on weapons lubrication specifically for cold weather operations. This lube is not only best for cold weather use, but the better news is that it does not evaporate or burn off during high heat, which I take to mean repeated firing of the weapon.

The maintenance instruction guide provided with each tube of FrogLube is the most comprehensive I have ever see for a weapon lubricant. It covers prep, solvent use, rifle bore cleaning, CLP (cleaning, lubricant, and protectant) application, cold temperature use, use in hot and dusty conditions, storage, and long-term care. Everything you need to know about lubing your weapons.

So far in small, judicious applications of this lube on a couple new firearms that were new out of the box, I can already tell there is a slickness to this lube I have not noticed in conventional lubricants for guns. The slides work noticeably easier. I anticipate good results at the range, too, after some heated use of these firearms.

I am also looking forward to seeing if FrogLube will be the solution to one particular AR rifle I have in .308. I have run this rifle wet and dry, and the bolt hangs up in either case, worse when run dry. I will report on this trial later.

FrogLube has been around for a while so it is not really new. I recommend you try it. Contact FrogLube at www.froglube.com.

GRADUATE TO ONE-PIECE CLEANING RODS

Sometimes in life you have to learn lessons the hard way. For years and years I cleaned all of my hunting rifles, shotguns, rimfire plinking rifles, and survival prepping/varmint ARs with two- or even three-piece jointed cleaning rods. I bet you have too, and learned just like I did the fallacies of using these sectioned rods for gun-barrel cleaning.

First on the list of bad characteristics of the jointed cleaning rods is the inherent weakness they have in driving a cleaning patch from the breech end down to the muzzle. On more than one occasion I have had a jointed rod seize up inside the barrel. Either the patch was too large or some hydrologic phenomenon with the cleaning solution caused the rod to lock down.

Then, the force needed to either reverse the direction of the rod or push it out the other end could not be mustered with regular arm-driving pressure. This resulted in having to apply short strokes of a rubber mallet on the hammer to get the wet patch out the muzzle end. I don't even want to know if the jointed edges of the rod sections rubbed up against the barrel rifling, causing damage.

Twice I can recall trying to push a two-piece rod through a tight bore when the rod simply collapsed and broke off at the threaded coupling point. This action resulted in using lock-down pliers to extract the piece of the rod already inserted partway into the bore.

So finally I broke down and bought a high-quality one-piece, heavy-duty, coated stainless steel rifle cleaning rod and my troubles ended. With a one-piece rod you can carefully apply sufficient torque to push a tight patch clear of the muzzle. Now I have bought a second one-piece rod to handle the .22 caliber bores.

If you are still using the old, weak, cheap, screw-together gun-cleaning rods, then do yourself a favor and trash them in favor of good one-piece rods. You'll see the difference the first time you use one.

KEEP A GUNSMITHING SCREWDRIVER SET ON HAND

Working on a firearm is nothing like working on a car engine or lawnmower. Gun screws are typically finely made precision parts. Therefore, they demand precision tools for working on them. If you plan to tinker on your firearms, then the first thing you need to buy is a high-quality gunsmithing screwdriver set.

How many times have you gone to a gun show or gun dealer in search of a used gun for this or that shooting task and seen the abuse some of them have received? I just cringe when I see gun screws with the head slots twisted out or buggered up to the point that they cannot even be extracted. This tells me the gun was not taken care of and it is not a gun I want to buy.

Precision screws and other assembly attachment parts demand well-fitting tools. Screwdriver sets manufactured specifically to use on guns are typically high quality, well-made, and designed to fit exactly into screw slots or hex heads, or other specialty slots like a Torx® head.

A good set of gunsmithing screwdrivers will actually consist of a handle or two, often one short stubby handle and one longer one into which a wide variety of tool bits will fit. Often the screwdriver handle head is magnetized to retain the screw bit in place. These sets will include a wide variety of both Phillips, square drivers, slotted bits, Allen hex bits, and the Torx bits.

There are many such gunsmithing screwdriver sets on the market, but among the best is the Chapman set that comes complete in a covered box with a handle, a snap-in screwdriver shaft, and multiple bits. This set also includes a mini-ratchet that allows for extra torque and speed. Chapman also makes a supplemental set or Adapter Pack that contains 18 additional Allen hex and metric hex-type adapters.

Another extremely well-made and high-quality set of gunsmithing screwdrivers is the set offered by Brownell's, the professional gunsmithing supply outlet in Iowa. Their set of drivers comes with two handles and every possible size of flat head, Phillips, and other adapter heads all set inside a snap-closed heavy plastic toolbox case. This set is first rate. If you are going to work on guns even as a hobby, get a good gunsmithing screwdriver set.

ABOUT THE AUTHOR

John J. Woods, Ph.D. has a B.S. degree in Biology, a Master's Degree in Wildlife Science, and a Ph.D. in Industrial Psychology. He has been a Certified Hunter Education Instructor (Missouri and Mississippi) since 1980, is an NRA Certified Firearms Appraiser and a Personal Firearms Counselor. Most recently he was Vice President of Economic Development and Training at Eagle Ridge Conference and Training Center, Hinds Community College, Mississippi's largest community college.

John has been an award-winning author of outdoors material for over 37 years. He's served as a masthead editor, contributing editor, field editor, columnist, freelance writer, and industry consultant, with over 3,000 articles and columns published. During that time he has contributed to eight books, including The American Deer Camp and The Magnolia Club.

He's a proud member of POWR, NRA Life, North American Hunting Club Life, Mississippi Wildlife Federation Life & Board of Directors, Buckmasters Life, and Quality Deer Management Association.

John specializes in hunting (especially white-tailed deer and turkey), salt and freshwater fishing, conservation, environment, product reviews, travel and hunting destinations, guns and gear, ARs, and ATVs.

PERMUTED PRESS
needs *you* to help

SPREAD (THE) INFECTION

FOLLOW US!

- Facebook.com/PermutedPress
- Twitter.com/PermutedPress

REVIEW US!

Wherever you buy our book, they can be reviewed! We want to know what you like!

GET INFECTED!

Sign up for our mailing list at PermutedPress.com

KING ARTHUR AND THE KNIGHTS OF THE ROUND TABLE HAVE BEEN REBORN TO SAVE THE WORLD FROM THE CLUTCHES OF MORGANA WHILE SHE PROPELS OUR MODERN WORLD INTO THE MIDDLE AGES.

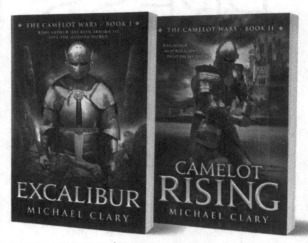

EAN 9781618685018 $15.99   **EAN** 9781682611562 $15.99

Morgana's first attack came in a red fog that wiped out all modern technology. The entire planet was pushed back into the middle ages. The world descended into chaos.

But hope is not yet lost— King Arthur, Merlin, and the Knights of the Round Table have been reborn.

THE ULTIMATE PREPPER'S ADVENTURE.
THE JOURNEY BEGINS HERE!

EAN 9781682611654 $9.99   EAN 9781618687371 $9.99   EAN 9781618687395 $9.99

The long-predicted Coronal Mass Ejection has finally hit the Earth, virtually destroying civilization. Nathan Owens has been prepping for a disaster like this for years, but now he's a thousand miles away from his family and his refuge. He'll have to employ all his hard-won survivalist skills to save his current community, before he begins his long journey through doomsday to get back home.

THE MORNINGSTAR STRAIN HAS BEEN LET LOOSE—IS THERE ANY WAY TO STOP IT?

An industrial accident unleashes some of the Morningstar Strain. The

EAN 9781618686497 $16.00

doctor who discovered the strain and her assistant will have to fight their way through Sprinters and Shamblers to save themselves, the vaccine, and the base. Then they discover that it wasn't an accident at all—somebody inside the facility did it on purpose. The war with the RSA and the infected is far from over.

This is the fourth book in Z.A. Recht's The Morningstar Strain series, written by Brad Munson.

PERMUTED PRESS

GATHERED TOGETHER AT LAST, THREE TALES OF FANTASY CENTERING AROUND THE MYSTERIOUS CITY OF SHADOWS...ALSO KNOWN AS CHICAGO.

EAN 9781682612286 $9.99   **EAN** 9781618684639 $5.99   **EAN** 9781618684899 $5.99

From *The New York Times* and *USA Today* bestselling author Richard A. Knaak comes three tales from Chicago, the City of Shadows. Enter the world of the Grey–the creatures that live at the edge of our imagination and seek to be real. Follow the quest of a wizard seeking escape from the centuries-long haunting of a gargoyle. Behold the coming of the end of the world as the Dutchman arrives.

Enter the City of Shadows.

PERMUTED PRESS

WE CAN'T GUARANTEE THIS GUIDE WILL SAVE YOUR LIFE. BUT WE CAN GUARANTEE IT WILL KEEP YOU SMILING WHILE THE LIVING DEAD ARE CHOWING DOWN ON YOU.

EAN 9781618686695 $9.99

This is the only tool you need to survive the zombie apocalypse.

OK, that's not really true. But when the SHTF, you're going to want a survival guide that's not just geared toward day-to-day survival. You'll need one that addresses the essential skills for true nourishment of the human spirit. Living through the end of the world isn't worth a damn unless you can enjoy yourself in any way you want. (Except, of course, for anything having to do with abuse. We could never condone such things. At least the publisher's lawyers say we can't.)

PERMUTED PRESS